John Disturnell

The Great Lakes

Inland Seas of America

John Disturnell

The Great Lakes
Inland Seas of America

ISBN/EAN: 9783743442689

Manufactured in Europe, USA, Canada, Australia, Japa

Cover: Foto ©berggeist007 / pixelio.de

Manufactured and distributed by brebook publishing software (www.brebook.com)

John Disturnell

The Great Lakes

OR

INLAND SEAS OF AMERICA;

EMBRACING A FULL DESCRIPTION OF

LAKES SUPERIOR, HURON, MICHIGAN, ERIE, AND ONTARIO;

RIVERS ST. MARY, ST. CLAIR, DETROIT, NIAGARA, AND ST. LAWRENCE:

LAKE WINNIPEG, ETC.:

TOGETHER WITH THE

COMMERCE OF THE LAKES,

AND

TRIPS THROUGH THE LAKES:

GIVING A DESCRIPTION OF CITIES, TOWNS, ETC.

FORMING ALTOGETHER

A Complete Guide

FOR THE PLEASURE TRAVELLER AND EMIGRANT.

With Map and Embellishments.

COMPILED BY J. DISTURNELL,

AUTHOR OF THE "PICTURESQUE TOURIST," ETC.

NEW YORK:

PUBLISHED BY CHARLES SCRIBNER,

NO. 124 GRAND STREET,

1863.

PREFACE.

In presenting to the Public the present volume, entitled "THE GREAT LAKES, or 'INLAND SEAS' of AMERICA," embracing the Magnitude of the Lakes—Commerce of the Lakes—Trip through the Lakes—Route from Lake Superior to Lake Winnipeg—Tables of Distances, etc., together with a Description of Niagara Falls and the St. Lawrence River, the Compiler wishes to return his sincere thanks for the liberal patronage and the many kind favors received from those who have doubly assisted him, by contributing reliable and useful information in regard to the many interesting localities in which the Great Valley of the Lakes and of the St. Lawrence abounds, affording altogether many new and interesting facts of great importance to the Tourist, who may wish to visit the Inland Seas of America for health or pleasure; the tour being one of the most healthy, picturesque, and wonderful—when viewed as a whole, from Lake Winnipeg to Lake Superior, and thence to the Gulf of St. Lawrence—on the face of the globe.

In the arrangement and compilation of this work every attempt has been made to render the information it contains concise and truthful. The Magnitude and also the Commerce of the Lakes are themes of great interest, they now being whitened by a large fleet of sail-vessels, in addition to the swift steamers and propellers which plough the waters of these Great Lakes, transporting annually large numbers of passengers, and an immense amount of agricultural and mineral products, to and from the different ports.

Lake Superior, the *Ultima Thule* of many travellers, can now be easily reached by lines of steamers starting from Cleveland and Detroit, running through Lake Huron and the St. Mary's River; or from Chicago and Milwaukee, passing through Lake Michigan and the Straits of Mackinac. All these routes are fully described, in connection with the Collingwood Route, and Route to Green Bay; affording altogether ample and cheap opportunities to visit every portion of the Upper Lakes and their adjacent shores.

The most noted places of Resort are Mackinac, Saut Ste. Marie, Munising, near the Pictured Rocks, Marquette, Houghton, Copper Harbor, Ontonagon, Bayfield, and Superior City. If to these should be added a Trip to the North Shore of Canada, visiting Michipicoten Island, Fort William, and other interesting localities—passing Isle Royale, attached to the State of Michigan—the intelligent Tourist would see new wonders of almost indescribable interest, while inhaling the pure atmosphere of this whole region of Lake Country. Among the Mineral Regions may be found objects of interest sufficient to induce the tourist to spend some weeks or months in exploration, hunting, and fishing; and if, added to this, should be included a journey to the Upper Mississippi Valley, or Lake Winnipeg, an entire season could be profitably employed.

For a full description of the Lower St. Lawrence, Lake Champlain, Saguenay River, etc., the Traveller is referred to the "PICTURESQUE TOURIST," issued a few years since.   J. D.

NEW YORK, *July*, 1863.

CONTENTS·

PART I.

PART II.

PART III.

PART IV.

Route from Lake Superior to Lake Winnipeg.

PART V.

PART VI.

List of Embellishments.

TABLE OF DISTANCES,

From Boston, New York, Philadelphia and Baltimore, to Niagara Falls, Buffalo, Cleveland. &c.

		Miles.
1.	BOSTON to ALBANY, via *Western Railroad* of Massachusetts,...........	200
	ALBANY to NIAGARA FALLS, via *New York Central Railroad*,...,......306—506	
	NIAGARA FALLS to DETROIT, Mich., via *Great Western Railway of Canada*,. 230—736	
2.	NEW YORK to ALBANY, via *Hudson River Railroad*,...................	145
	ALBANY to BUFFALO, via *New York Central Railroad*,................298—443	
3.	NEW YORK to BUFFALO, via *Erie Railway*,...........................	432
	BUFFALO to CLEVELAND, Ohio, via *Lake Shore Railroad*,.............183—615	
4.	NEW YORK to HARRISBURG, via *N. J. Central Railway*..............	182
	HARRISBURG to PITTSBURGH, via *Pennsylvania Central Railroad*,........249—431	
	PITTSBURGH to CLEVELAND, via *Pittsburgh and Cleveland Railway*,......150—581	
5.	PHILADELHHIA to PITTSBURGH, via *Pennsylvania Central Railroad*,......	356
	PITTSBURGH to CLEVELAND, via *Pittsburgh and Cleveland Railroad*,......150—506	
6.	PHILADELPHIA to ELMIRA, via *Philadelphia and Elmira Railway*,.......	275
	ELMIRA to BUFFALO, via *Erie Railway*,............................159—434	
7.	BALTIMORE to HARRISBURG, via *Northern Central Railway*,.............	85
	HARRISBURG to ELMIRA, N. Y. " " " ....171—236	
	ELMIRA to BUFFALO, via *Erie Railway*,.............................159—415	
8.	BALTIMORE to PITTSBURGH, via *Pennsylvania Central Railroad*,..........	334
	PITTSBURGH to CLEVELAND, Ohio, via *Pittsburgh and Cleveland Railroad*,. 150—484	
	CLEVELAND to DETROIT, Mich., via *Steamboat Route*,..................120—604	

BOSTON to NEW YORK, *Railroad Route*,............................	236
NEW YORK to PHILADELPHIA, *Railroad Route*,.....................	90—326
PHILADELPHIA to BALTIMORE, " "	98—424
BALTIMORE to WASHINGTON, " "	40—464

RAILROAD AND STEAMBOAT ROUTE,

FROM NEW YORK TO NIAGARA FALLS AND TORONTO, C. W., LEAVING NEW YORK
AT SIX P. M. BY STEAMER.

Stations, etc.	Miles.	Usual Time. H. M.
NEW YORK..........................	0	
ALBANY, (Steamer).....................	145	12 00
Schenectady, (Railroad).................	162	13 00
Utica, "	240	16 00
Rome, "	254	16 30
Syracuse, "	293	18 00
ROCHESTER, (St. to Toronto).............	374	22 45
Lockport, (Railroad).............	430	25 00
SUSPENSION BRIDGE, "	448	26 00
LEWISTON, "	452	
TORONTO, (Steamer).............	494	30 00

RAILROAD AND STEAMBOAT ROUTE,

FROM NEW YORK TO OSWEGO, TORONTO, ETC., LEAVING NEW YORK AT 7 & 10
A. M., AND 5 P. M., BY HUDSON RIVER RAILROAD.

Stations, etc.	Miles.	Usual Time. H. M.
NEW YORK..........................	0	
Poughkeepsie, (Railroad)...............	75	2 40
Hudson, "	116	4 00
ALBANY, "	144	6 00
Schenectady, "	162	6 00
Utica, "	240	8 30
Rome, "	254	9 00
Syracuse, "	293	10 30
OSWEGO. "	323	13 00
LEWISTON, (Steamer 140 m.).............	408	
TORONTO, (Steamer 150 m.).............	478	27 00

NOTE.—Passengers by continuing on by Railroad from Syracuse, via Rochester and
Lockport, will arrive at Suspension Bridge, 418 miles, in sixteen hours after leaving
New York, stop at Niagara Falls if desired, and reach Toronto by Railroad, via
Hamilton, C. W., 81 miles farther; making the total distance from New York to
Toronto by Railroad, via Suspension Bridge, 529 miles.

RAILROAD AND STEAMBOAT ROUTE

From New York to Lake Superior.

Stopping Places		Total Miles.		Usual Time. Hours.
NEW YORK to ALBANY, by *Railroad*..			145	5
ALBANY TO BUFFALO, "		298	443	10
ALBANY to NIAGARA FALLS, "		304		
BUFFALO to CLEVELAND, Ohio, "		183	626	7
CLEVELAND to DETROIT, by *Steamboat*,.		120	746	10
DETROIT to PORT HURON, "		73	819	6
PORT HURON to SAUT STE. MARIE " ..		277	1,096	24
SAUT STE. MARIE to MARQUETTE " ..		160	1,256	14
MARQUETTE to ONTONAGON, " ..		226	1,482	20
ONTONAGON to BAYFIELD, " ..		78	1,560	7
BAYFIELD to SUPERIOR CITY, " ..		80	1,640	7

Total Running Time, 4 days and 14 hours.

USUAL FARE from New York to Buffalo.....................	$ 9	35
" " New York to Cleveland, O...............	14	35
" " New York to Detroit, Mich............	16	35
Detroit to Lake Superior and Return.................	25	00

RAPIDS OF THE ST. LAWRENCE RIVER.

The Rapids of the St. Lawrence, in connection with the "*Thousand Islands*," form the most remarkable feature of this truly noble stream. The "Thousand Islands" are situated near the foot of Lake Ontario, where the St. Lawrence proper commences. Here are found delightful resorts for those fond of fishing and hunting, surrounded by scenery of the most enchanting character.

The fall in the St. Lawrence river, between Ogdensburgh and Montreal, a distance of 120 miles exceeds 200 feet. The rapids encountered are the *Gallop Rapids; Rapid Plat; Long Saut Rapids*, (descent 48 feet.) The *Coteau Rapids, Cedar Rapids*, and *Cascade Rapids*, have a descent of 82 feet; in the distance of 11 miles. The *La Chine Rapids*, the last formidable rapids which impede navigation, has a descent of 45 feet.

The descent of these rapids by steamers is perfectly safe, affording the most exciting and grand excursion imaginable, In ascending the stream steamers pass through the *St. Lawrence Canal;* total length about 40 miles. *See Engraving,* page 162.

MAGNITUDE OF THE LAKES, OR "INLAND SEAS."

OTHING but a voyage over all of the great bodies of water forming the "INLAND SEAS," can furnish the tourist, or scientific explorer, a just idea of the extent, depth, and clearness of the waters of the Great Lakes of America, together with the healthy influence, fertility, and romantic beauty of the numerous islands, and surrounding shores, forming a circuit of about 4,000 miles, with an area of 90,000 square miles, or about twice the extent of the State of New York—extending through eight degrees of latitude, and sixteen degrees of longitude—this region embracing the entire north half of the 'temperate zone, where the purity of the atmosphere vies with the purity of these extensive waters, or "Inland Seas," being connected by navigable rivers or straits.

The States, washed by the Great Lakes, are New York, Pennsylvania, Ohio, Michigan, Indiana, Illinois, Wisconsin, Minnesota, and Canada West—the boundary line between the United States and the British Possessions running through the centre of Lakes Superior, Huron, St. Clair, Erie, and Ontario, together with the connecting rivers or straits, and down the St. Lawrence River to the 45th parallel of latitude. From thence the St. Lawrence flows in a northeast direction through Canada into the Gulf of St. Lawrence. The romantic beauty of the rapids of this noble stream, and its majestic flow through a healthy and rich section of country, is unsurpassed for grand lake and river scenery.

Lake Superior, the largest of the Inland Seas, lying between 46° 30' and 49° north latitude, and between 84° 30' and 92° 30' west longitude from Greenwich, is situated at a height of 600 feet above the Gulf of St. Lawrence, from which it is distant about 1,500 miles by the course of its outlet and the St. Lawrence river. It is 460 miles long from east to west, and 170 miles broad in its widest part, with an average breadth of 85 miles; the entire circuit being about 1,200 miles. It is 800 feet in greatest depth, extending 200 feet below the level of the ocean. Estimated area, 31,500 square miles, being by far the largest body of fresh water on the face of the globe—celebrated alike for its sparkling purity, romantic scenery, and healthy influence of its surrounding climate. About one hundred rivers and creeks are said to flow into the lake, the greatest part being small streams, and but few navigable except for canoes, owing to numerous falls and rapids. It discharges its waters eastward, by the strait, or river *St. Mary,* 60 miles long, into Lake Huron, which lies 26 feet below, there being about 20 feet descent at the Saut Ste Marie, which is overcome by means of two locks and a ship canal. Its outlet, is a most lovely and romantic stream, embosoming a number of large and fertile islands, covered with a rich foliage.

Lake Michigan, lying 576 ft. above the sea, is 320 miles long, 85 miles broad, and 700 feet deep; area, 22,000 square miles. This lake lies wholly within the confines of the United States. It presents a large expanse of water, with but few islands, except near its entrance into the Straits of Mackinac, through which it discharges its surplus waters. The strait is 30 or 40 miles in length, and discharges its accumulated waters into Lake Huron, on nearly a level with Lake Michigan. At the north end of the lake, and in the Straits, are several large and romantic islands, affording delightful resorts.

Green Bay, a most beautiful expanse of water, containing several small islands, lies at about the same elevation as Lake Michigan; it is 100 miles long, 20 miles broad, and 60 feet deep; area, 2,000 square miles. This is a remarkably pure body of water, presenting lovely shores, surrounded by a fruitful and healthy section of country.

Lake Huron, lying at a height of 574 feet above the sea, is 250 miles long, 100 miles broad, and 750 feet greatest depth; area, 21,000 square miles. This lake is almost entirely free of islands, presenting a large expanse of pure water. Its most remarkable feature is Saginaw Bay, lying on its western border. The waters of this lake are now whitened by the sails of commerce, it being the great thoroughfare to and from Lakes Michigan and Superior.

Georgian Bay, lying northeast of Lake Huron, and of the same altitude, being separated by islands and headlands, lies wholly within the confines of Canada. It is 140 miles long, 55 miles broad, and 500 feet in depth; area, 5,000 square miles. In the *North Channel*, which communicates with St. Mary's River, and in Georgian Bay, are innumerable islands and islets, forming an interesting and romantic feature to this pure body of water. All the above bodies of water, into which

are discharged a great number of streams, find an outlet by the River *St. Clair*, commencing at the foot of Lake Huron, where it has only a width of 1,000 feet, and a depth of from 20 to 60 feet, flowing with a rapid current downward, 38 miles, into

Lake St. Clair, which is 25 miles long and about as many broad, with a small depth of water; the most difficult navigation being encountered in passing over "*St. Clair Flats,*" where only about 12 feet of water is afforded. *Detroit River,* 27 miles in length, is the recipient of all the above waters, flowing southward through a fine section of country into

Lake Erie, the *fourth* great lake of this immense chain. This latter lake again, at an elevation above the sea of 564 feet, 250 miles long, 60 miles broad, and 204 feet at its greatest depth, but, on an average, considerably less than 100 feet deep, discharges its surplus waters by the Niagara River and Falls, into Lake Ontario, 330 feet below; 51 feet of this descent being in the rapids immediately above the Falls, 160 feet at the Falls themselves, and the rest chiefly in the rapids between the Falls and the mouth of the river, 35 miles below Lake Erie. This is comparatively a shallow body of water; and the relative depths of the great series of lakes may be illustrated by saying, that the surplus waters poured from the vast *basins* of Superior, Michigan, and Huron, flow across the *plate* of Erie into the deep *bowl* of Ontario. Lake Erie is reputed to be the only one of the series in which any current is perceptible. The fact, if it is one, is usually ascribed to its shallowness; but the vast volume of its outlet—the Niagara River—with its strong current, is a much more favorable cause than the small depth of its water, which may be far more appropriately adduced as the reason why the navigation is obstructed by *ice* much more than either of the other great lakes.

The ascertained temperature in the middle of Lake Erie, August, 1845, was temperature of air 76° Fahrenheit, at noon—water at surface 73°—at bottom 53°.

Lake Ontario, the *fifth* and last of the Great Lakes of America, is elevated 234 feet above tide-water at Three Rivers on the St. Lawrence; it is 180 miles long, 60 miles broad, 600 feet deep.

Thus *basin* succeeds *basin*, like the locks of a great canal, the whole length of waters from Lake Superior to the Gulf of St. Lawrence being rendered navigable for vessels of a large class by means of the Welland and St. Lawrence canals—thus enabling a loaded vessel to ascend or descend 600 feet above the level of the ocean, or tide-water. Of these five great lakes, Lake Superior has by far the largest area, and Lake Ontario has the least, having a surface only about one-fifth of that of Lake Superior, and being somewhat less in area than Lake Erie, although not much less, if any, in the circuit of its shores. Lake Ontario is the safest body of water for navigation, and Lake Erie the most dangerous. The lakes of greatest interest to the tourist or scientific traveler are Ontario, Huron, together with Georgian Bay and North Channel, and Lake Superior. The many picturesque islands and headlands, together with the pure dark green waters of the Upper Lake, form a most lovely contrast during the summer and autumn months.

The altitude of the land which forms the water-shed of the *Upper Lakes* does not exceed from 600 to 2,500 feet above the level of the ocean, while the altitude of the land which forms the water-shed of Lake Champlain and the lower tributaries of the St. Lawrence River rises from 4,000 to 5,000 above the level of the sea or tide-water, in the States of Vermont and New York.

The divide which separates the waters of the Gulf of Mexico, from those flowing northeast into the St. Lawrence, do not in some places exceed ten or twenty feet above the level of Lakes Michigan and Superior; in fact, it is said that Lake Michigan, when under the influence of high water and a strong northerly wind, discharges some of its surplus waters into the Illinois River, and thence into the Mississippi and Gulf of Mexico—so low is the divide at its southern terminus.

When we consider the magnitude of these Great Lakes, the largest body of fresh water on the globe, being connected by navigable Straits, or canals, we may quote with emphasis the words of an English writer: "How little are they aware, in Europe, of the extent of commerce upon these 'Inland Seas,' whose coasts are now lined with flourishing towns and cities; whose waters are plowed with magnificent steamers, and hundreds of vessels crowded with merchandise! Even the Americans themselves are not fully aware of the rising importance of these great lakes, as connected with the Far West.

TRIBUTARIES OF THE GREAT LAKES AND ST. LAWRENCE RIVER.

Unlike the tributaries of the Mississippi, the streams falling into the Great Lakes or the St. Lawrence River are mostly rapid, and navigable only for a short distance from their mouths.

The following are the principal Rivers that are navigable for any considerable length:

AMERICAN SIDE.   Miles.

St. Louis River, Min..............	Superior to Fond du Lac.............	20
Fox, or Neenah, Wis...,..............	Green Bay to Lake Winnebago*.....	36
St. Joseph, Mich.....................	St. Joseph to Niles.................	26
Grand River, "	Grand Haven to Grand Rapids......	40
Muskegon, "	Muskegon to Newaygo.............	40
Saginaw "	Saginaw Bay to Upper Saginaw.....	26
Maumee, Ohio	Maumee Bay to Perrysburgh........	18
Genesee, N. Y.....................	Charlotte to Rochester.............	6

CANADIAN SIDE.   Miles.

Thames...........................	Lake St. Clair to Chatham...........	24
Ottawa	La Chine to Carillon...............	40
"	(By means of locks to Ottawa City)†...	70
Richelieu or Sorel....................	Sorel to Lake Champlain (by locks)	75
Saguenay	Tadusac to Chicoutimi...........	70
	(thence to Lake St. John, 50 m.)	

LAKE AND RIVER NAVIGATION,

FROM FOND DU LAC, LAKE SUPERIOR, TO THE GULF OF ST. LAWRENCE.

LAKES, RIVERS, ETC.	Length in miles.	Greatest breadth.	Av. breadth.	Depth in feet.	El. above sea.
Superior......................	460	170	85	800	600 ft.
St. Mary's River................	60	5	2	10 to 100	
Michigan......................	320	85	58	700	576 "
Green Bay.....................	100	25	18	100	576 "
Strait of Mackinac..............	40	20	10	20 to 200	575 "
Huron.......................	250	100	70	700	574 "
North Channel.................	150	20	10	20 to 200	574 "
Georgian Bay..................	140	65	40	500	574 "
St. Clair River.................	38	1¼	*1	20 to 60	
Lake St. Clair*............. ..	25	25	18	10 to 20	568 "
Detroit River..................	27	3	1	10 to 60	
Erie.........................	250	70	40	200	564 "
Niagara River.................	35	3	.1		
Ontario......................	180	58	40	600	234 "
St. Lawrence River.............	760	100	2		
Lake St. Francis, foot Long Saut...			4		142 "
Lake St. Louis, foot Cascade Rapids			5		58 "
At Montreal...................			3		13 "
Lake St. Peter.................			12		6 "
Tide-water at Three Rivers.......			1		0 "
At Quebec....................			1		0 "

Total miles navigation........ 2,835

* By means of 17 locks, overcoming an elevation of 170 feet.
† The navigation for steamers extends 150 miles above Ottawa City, by means of portages and locks.
* The St. Clair Flats, which have to be passed by all large steamers and sail vessels running from Lake Erie to the Upper Lakes, now affords twelve feet of water, the ship channel being very narrow and winding, where stands a light and buoys to guide the mariner.

ALTITUDE OF VARIOUS POINTS ON THE SHORES OF LAKE SUPERIOR.

LOCALITIES.	Above Lake Superior.	Above the Sea.
Lake Superior.....................................	000 feet.	600 feet.
Point Iroquois, South Shore..................	350 "	950 "
Gros Cap, C. W., North Shore................	700 "	1,300 "
Grand Sable,　　South Shore	345 "	945 "
Pictured Rocks,　　"　　.............\..........	200 "	800 "
Iron Mountains,　　"　　....................	850 "	1,450 "
Quincy Copper Mine, Portage Lake.............	550 "	1,150 "
Mount Houghton, near Keweenaw Point........	1,000 "	1,600 "
Porcupine Mountains, South Shore.............	1,380 "	1,980 "
Isle Royale, Michigan.....	300 "	900 "
Minnesota Mountains (estimated)...............	1,200 "	1,800 "
Michipicoten Island, C. W....................	800 "	1,400 "
Pie Island,　　　　"　　....................	760 "	1,360 "
St. Ignace (estimated)　"　....................	1,200 "	1,800 "
McKay's Mountain,　　"　....................	1,000 "	1,600 "
Thunder Cape,　　　　"　....................	1,350 "	1,950 "

TOPOGRAPHY AND METEOROLOGY.

"The mountains of the region along the south shore of Lake Superior, consist of two granite belts in the northwest, the *Huron Mountains* to the southward, a trap range starting from the head of Keweenaw Point, and running west and southwest into Wisconsin, the *Porcupine Mountains*, and the detrital rocks. The Huron Mountains in places attain an elevation of 1,200 feet above the Lake. The highest elevation attained by the Porcupine Mountains is 1,380 feet.

"Meteorological observations were instituted by order of the Government at three military posts in the District, viz.: Forts Wilkins (Copper Harbor), Brady, and Mackinac. From these observations it appears that the mean annual temperature of Fort Brady is about one degree lower than that of Fort Wilkins, although the latter post is nearly a degree further north. This difference arises from the insular position of Keweenaw Point, which is surrounded on three sides by water. The climate at Fort Brady, dur-

ing the whole season, corresponds in a remarkable degree with that of St. Petersburg. The temperature of the region is very favorable to the growth of cereals. The annual ratio of fair days at Fort Brady is 168; of cloudy days, 77; rainy days, 71; snowy days, 47.

"The temperature of the water of Lake Superior during the summer, a fathom or two below the surface, is but a few degrees above the freezing point. In the western portion, the water is much colder than in the eastern—the surface flow becoming warmer as it advances toward the outlet. The mirage which frequently occurs, is occasioned by the difference between the temperature of the air and the Lake. Great difficulties are experienced from this cause in making astronomical observations.

"Auroras, even in midsummer, are of frequent occurrence, and exhibit a brilliancy rarely observed in lower latitudes."
—*Foster & Whitney's Report.*

2

THE UPPER LAKES, OR "INLAND SEA," OF AMERICA.

This appellation applies to Lakes Huron, Michigan, and Superior, including Green Bay, lying within the confines of the United States, and Georgian Bay, which lies entirely in Canada.

These bodies of water embrace an area of about 75,000 square miles, and, as a whole, are deserving of the name of the 'INLAND SEA,' being closely connected by straits or water-courses, navigable for the largest class of steamers or sail vessels. The shores, although not elevated, are bold, and free from marsh or swampy lands, presenting one clean range of coast for about 3,000 miles.

By a late decision of the Supreme Court of the U. States, the Upper Lakes including Lake Erie, with their connecting waters, were declared to be *seas*, commercially and legally. Congress, under this decision, is empowered to improve the harbors of the lakes and the connecting straits, precisely as it has power to do the same on the seaboard. This will probably lead to a vigorous policy in the maintenance of Federal authority, both in improving the harbors, and making provision for the safety of commerce, and protection of life, as well as guarding against foreign invasion. The only fortification of importance that is garrisoned is *Fort Mackinac*, guarding the passage through the Straits of Mackinac.

The islands of these lakes are numerous, particularly in the Straits of Mackinac, and in Georgian Bay, retaining the same bold and virgin appearance as the mainland; most of them are fertile and susceptible of high cultivation, although, as yet, but few are inhabited to any considerable extent.

The dark green waters of the Upper Lakes, when agitated by a storm, or the motion of a passing steamer, presents a brilliancy peculiar only to these transparent waters—they then assume the admixture of white foam, with a lively green tinge, assuming a crystal-like appearance. In this pure water, the *white fish*, and other species of the finny tribe, delight to gambol, affording the sportsman and epicurean untold pleasure, which is well described in the following poem:

THE WHITE FISH.

HENRY R. SCHOOLCRAFT, in his poem. "THE WHITE FISH," says:

" All friends to good living by tureen and dish,
Concur in exulting this prince of a fish ;
So fine in a platter, so tempting a fry,
So rich on a gridiron, so sweet in a pie ;
That even before it the salmon must fall,
And that mighty *bonne-bouche*, of the land-
     beaver's tail.

*   *   *   *

'Tis a morsel alike for the gourmand or faster,
While, white as a tablet of pure alabaster !
Its beauty or flavor no person can doubt,
When seen in the water or tasted without ;
And all the dispute that opinion ere makes
Of this king of lake fishes, this '*deer of the
     lakes*,'*
Regard not its choiceness to ponder or sap,
But the best mode of dressing and serving it up.

*   *   *

Here too, might a fancy to descant inclined,
Contemplate the love that pertains to the kind,
And bring up the red man, in fanciful strains,
To prove its creation from feminine brains."†

---

* A translation of *Ad-dik-keem-maig*, the Indian name for this fish.
† *Vide* " Indian Tales and Legends."

FISH OF THE UPPER LAKES.

"The numbers, varieties, and excellent quality of lake fish are worthy of notice. It is believed that no fresh waters known can, in any respect, bear comparison. They are, with some exceptions, of the same kind in all the lakes. Those found in Lake Superior and the straits of St. Mary are of the best quality, owing to the cooler temperature of the water. Their quantities are surprising, and apparently so inexhaustible, as to warrant the belief that were a population of millions to inhabit the lake shore, they would furnish an ample supply of this article of food without any sensible diminution. There are several kinds found in Lake Superior, and some of the most delicious quality, that are not found in the lakes below, as the siskowit and muckwaw, which grow to the weight of eight or ten pounds. The salmon and some others are found in Ontario, but not above the Falls of Niagara.

"The following is a very partial list of a few of the prominent varieties: the white fish, Mackinac and salmon-trout, sturgeon, muscalunje, siskowit, pickerel, pike, perch, herring, white, black, and rock bass, cat, pout, eel-pout, bull-head, roach, sun-fish, dace, sucker, carp, mullet, bill-fish, sword-fish, bull-fish, stone-carrier, sheeps-head, gar, &c.

"The lamprey-eel is found in all, but the common eel is found in neither of the lakes, nor in any of their tributaries, except one. The weight to which some of these attain is not exceeded by the fish of any other inland fresh waters, except the Mississippi. *   *   *   *

"The fish seem to be more numerous some years than others, and likewise of better quality. The kinds best for pickling and export are the white fish, Mackinac and salmon trout, sturgeon, and pickerel. The fisheries at which these are caught are at Mackinac, at several points in each of the four straits, the southeast part of Lake Superior, Thunder Bay, Saginaw Bay, and Fort Gratiot near foot of Lake Huron. The sport of taking the brook trout, which are found in great abundance in the rapids at the Saut Ste Marie, and most all of the streams falling into the Upper Lakes, affords healthful amusement to hundreds of amateur fishermen during the summer and fall months. The modes of taking the different kinds of fish are in seines, dip-nets, and gill-nets, and the trout with hooks.

"Those engaged in catching fish in the Straits of Mackinac, are composed of Americans, Irish, French, half breeds, and Indians. Some are employed by capitalists, others have their own boats and nets. Each one is furnished with a boat, and from fifty to one hundred nets, requiring constantly two or three men for each boat, to run the different gangs of nets. The fish caught are principally white fish, with some trout. The demand for exportation increases every year, and although immense quantities are caught every season, still no diminution in their number is perceived.

"A fleet of two hundred fish-boats are engaged in and about the Straits, embracing, however, all the Beaver group. Each boat will average one barrel of fish per day during the fishing season.
*   *   *   *

"Ye, who are fond of sport and fun, who wish for wealth and strength; ye, who love angling; ye, who believe that God has given us a time to pray, a time to dance, &c., &c., go to these fishing-grounds, gain health and strength, and pull out Mackinac trout from 20 to 40 lbs. in weight. One hook and line has, in three to four hours, pulled out enough to fill three to four barrels of fish, without taking the sport into consideration.
"Yours, W. M. J.'

PHYSICAL CHARACTER OF THE MISSISSIPPI BASIN.

"The Valley of the Mississippi, bounded on the one hand by the Rocky Mountains, and on the other by the Alleghanies, embraces a drainage area of 1,244,000 square miles, which is more than one-half of the entire area of the United States. The Upper Mississippi Valley is composed of three subordinate basins, whose respective dimensions are as follows:

	Square miles.
The Ohio basin	214,000
The Upper Mississippi	169,000
The Missouri	518,000
Making a total of	901,000

Its navigable rivers are as follows:

	Miles.
Missouri, to near the Great Falls	3,150
Missouri, above Great Falls to Three Forks	150
Osage, to Osceola	200
Kansas	100
Big Sioux	75
Yellow-stone	800
Upper Mississippi, to St. Paul	658
St. Anthony, to Sauk Rapids	80
Above Little Falls, to Pokegima Falls	250
Minnesota, to Patterson's Rapids	295
St. Croix, to St. Croix Falls	60
Illinois, to La Salle	220
Ohio, to Pittsburgh	975
Monongahela, to Geneva(slack-water,4 locks)	91
Muskingum, to Dresden " 8 "	100
Green River to BowlingGreen " 5 "	186
Kentucky, to Brooklyn " 5 "	117
Kanawha, to Gauley Bridge	100
Wabash, to Lafayette	335
Salt, to Shepherdsville	30
Sandy, to Louisa	25
Tennessee, to Muscle Shoals	600
Cumberland, to Burkesville	370
Total navigation	8,967

NOTE.—Steamboats have ascended the Des Moines to Des Moines City, Iowa River to Iowa City, Cedar River to Cedar Rapids, and the Maquoketa to Maquoketa City, but only during temporary floods.

"It would thus appear that the internal navigation of the Upper Mississippi Valley is about 9,000 miles in extent; but, during the summer months, even through the main channels, it becomes precarious, and at times is practically suspended.

"The Mississippi Valley, viewed as a whole, may be regarded as one great plain between two diverging coast ranges, elevated from 400 to 800 feet above the sea. St. Paul, the head of the navigation of the Mississippi, is 800 feet above the ocean; Pittsburgh, at the junction of the Monongahela and Alleghany, forming the Ohio, 699 feet; Lake Superior on the north, 600 feet; but the water-shed on the west, at South Pass, rises to nearly 7,500 feet.

"It is traversed by no mountain ranges, but the surface swells into hills and ridges, and is diversified by forest and prairie. Leaving out the sterile portions west of the Missouri, the soil is almost uniformly fertile, easily cultivated, and yields an abundant return. The climate is healthy and invigorating, and altogether the region is the most attractive for immigration of any portion of the earth."

By means of a *Ship Canal*, connecting with the Illinois River, the waters of the Mississippi will be united with the waters of Lake Michigan—thus forming an uninterrupted navigation for armed vessels of a large class from the Gulf of Mexico to the Gulf of St. Lawrence, forming an inland navigation of about 3,000 miles—running through the sub-tropical and temperate zones, where nature is most lavish of her gifts.

PHYSICAL CHARACTER OF THE ST. LAWRENCE.

The sources of the Mississippi on the east interlock with those of the St. Lawrence, which, with its associated lakes and rivers, presents a system of water-communication of nearly equal extent and grandeur.

TABLE SHOWING THE DIMENSIONS OF THE
FIVE GREAT AMERICAN LAKES.

LAKES.	Greatest length. Miles.	Greatest breadth. Miles.	Height above sea. Feet.	Area in sq. miles.
Superior	460	170	600	81,500
Michigan.... ...	320	85	576	22,000
Huron	260	160	574	20,400
Erie	240	70	565	9,600
Ontario	180	60	232	6,500
Total......	1,460			90,000

"The entire area drained by these lakes
is estimated at 335,500 square miles, and
their shore lines are nearly 5,000 miles in
extent.

"These rivers are as diverse in character as in direction. The Mississippi is the
longer, but the St. Lawrence discharges
the greater volume of water; the one
abounds in difficult rapids, the other in
stupendous cataracts; the one is subject
to great fluctuations, the other preserves
an almost unvarying level; the waters of
the one are turbid, those of the other pos-
sess an almost crystal purity; the one
affords few lake-like expansions, the other
swells into vast inland seas. Both have
become the great highways of commerce,
enriching the regions through which they
flow, and supplying the inhabitants with
the varied products of distant climes."—
*Foster and Whitney's Report on Lake Supe-
rior.*

"The commerce of these lakes, whose
annual value reaches $450,000,000—more
than twice the external commerce of the
whole country—is carried on by a fleet of
1,643 vessels, of the following classes:—

	No.	Tonnage.	Value.
Steamers	143	54,522	$2,190,800
Propellers	254	70,253	3,573,800
Barks	74	84,203	942,900
Brigs................	85	24,831	523,900
Schooners..........	1,068	227,831	5,955,550
Sloops	16	667	12,770
Barges.............	8	8,719	17,000
Totals	1,643	413,026	$18,257,020

The following are the distances of some
of the commercial routes, taking Chicago
as the initial point:

Chicago to Mackinac (direct)	360 miles.
" " Fond du Lac Superior....	900 "
" " Georgian Bay............	650 "
" " Buffalo, N. Y...	950 "
" " Quebec	1,550 "
" " Gulf of St. Lawrence.....	1,950 "

PROGRESS OF DEVELOPMENT.

The first colony of English extraction,
planted in the territory of the Upper Mississippi, was in 1788—just seventy-five
years ago—at Marietta, within the present
limits of Ohio. This was the origin of
that spirit of colonization, which, within
the lifetime of many living men, has peopled this region with nine millions of human beings; has subdued and brought
under cultivation, an area greater than
that of all the cultivated lands of the British Empire; has connected the principal
commercial points with a net-work of railways more than eleven thousand miles in
extent; and has built up a domestic industry, the value of whose annual product
is in excess of three hundred and fifty
millions of dollars. Out of this territory
have been carved not less than nine States,
which are indissolubly linked together by
a similarity of conditions in soil and climate, and by the geographical features of
the country. They have already received
the appellation of the "FOOD-PRODUCING"
States—an appellation which they are
destined to retain for all time.

The rivers and the lakes, which water
this region, offer the most magnificent system of internal communication to be found
on the surface of the earth. No mountain
barriers interpose to divide the people into hostile clans, or divert the great currents of trade in their flow to the markets
of the world. With a soil sufficiently rich
in organic matter for fifty successive crops;
with almost boundless fields of coal, stored

away for future use; with vast deposits of the useful ores, and the precious metals, on the rim of the great basin; and with a climate most favorable to the development of human energy, it is impossible for the mind, even in its most daring speculations, to assign limits to the growth of the North-west. When all of these elements of wealth, now in a crude state, shall have been fully developed, there will be an exhibition of human power and greatness such as no other people ever attained.

Comparing the whole superficial content of these States with the portions cultivated, it will be seen that only about 16 per cent. of the surface has been subdued; and, if population and cultivation increase in the same ratio in the future as they have in the past, before the lapse of another decade there will be collected annually, on the borders of the Great Lakes, more than 200,000,000 bushels of cereals for exportation, giving employment to a fleet of more than 3,000 vessels, and requiring avenues of more than twice the capacity of existing ones.

A LUNAR TIDAL WAVE
In the North American Lakes.

Extract from a Paper read by Lt.-Col. Graham, *before the American Association for the Advancement of Science, August,* 1860.

"Much has been written, at various periods, on the fluctuations in the elevation of the surface waters of the great freshwater lakes of North America. Valuable and interesting memoirs have appeared from time to time in the American Journal of Science and Arts, published monthly at New Haven, Connecticut, within the last thirty years, on this subject, written by the late Brevet Brigadier-General Henry Whiting, of the U. S. Army, when a captain, by Major Lachlan, Charles Whittlesey, Esq., and others. The observations contained in their memoirs have, however, been directed chiefly to investigations of the extent of the secular and annual variations in elevation of the surfaces of these lakes.

"The learned Jesuit fathers of the time of Marquette, a period near two centuries ago, and at later periods the Baron de la Hontan, Charlevois, Carver, and others, noticed in their writings the changes of elevation, and some peculiar fluctuations which take place on these inland seas.

In the speculations indulged in by some of these writers a slight lunar tide is sometimes suspected, then again such an influence on the swelling and receding waters is doubted, and their disturbance is attributed to the varying courses and forces of the winds.

"But we have nowhere seen that any systematic course of observation was ever instituted and carried on by these early explorers, or by any of their successors who have mentioned the subject, giving the tidal readings at small enough intervals of time apart, and of long enough duration to develop the problem of a diurnal lunar tidal wave on these lakes. The general idea has undoubtedly been that no such lunar influence was here perceptible.

"In April, 1854, I was stationed at Chicago by the orders of the Government, and charged with the direction of the harbor improvements on Lake Michigan. In the latter part of August of that year, I caused to be erected at the east or lakeward extremity of the North harbor pier, a permanent tide-gauge for the purpose of making daily observations of the relative heights and fluctuations of the surface of this lake. The position thus chosen for the observations projects into the lake, entirely beyond the mouth of the Chicago River, and altogether out of the reach of any influence from the river current, upon the fluctuations of the tide-gauge. It was the fluctuations of the lake surface alone, that could affect the readings of the tide-gauge.

"On the first day of September, 1854, a course of observations was commenced on this tide-gauge, and continued at least once a day, until the 31st day of December, inclusive, 1858. During each of the first three winters a portion of the daily observations was lost, owing to the tide-gauge being frozen fast in its box, but they constituted only a small number in proportion to that embraced in the series. During the subsequent winters artificial means were resorted to, to prevent this freezing.

"These observations were instituted chiefly for the purpose of ascertaining with accuracy the amount of the annual, and also of the secular variation in the elevation of the lake surface, with a view to regulating the heights of break-waters and piers to be erected for the protection of vessels, and for improving the lake harbors."

After a series of close observations from 1854 to 1858, Lieut.-Colonel Graham observes:—

"The difference of elevation of the lake surface, between the periods of lunar low and lunar high-water at the mean spring tides is here shown to be two hundred and fifty-four thousandths (.254) of a foot, and the time of high-water at the full and change of the moon is shown to be thirty (30) minutes after the time of the moon's meridian transit.

"We, therefore, in accordance with custom in like cases, indicate as the *establishment* for the port of Chicago,

h.   m.
¼ Foot, 0 30.

"Although this knowledge may be of but small practical advantage to navigators, yet it may serve as a memorandum of a physical phenomenon whose existence has generally heretofore been either denied or doubted.

"We think it probable that, if the effect of unfavorable winds and all other extraneous forces which produce irregular oscillations in the elevation of the lake surface could be fully eliminated, a semi-diurnal lunar spring tide would be shown of as much as one-third of a foot for the periods of highest tides.

The time of low-water and the relative times of duration of the flood and ebb tides are given only approximately. The extreme rise of the tide being so little, the precise time of the change from ebb to flood, and hence the duration of the flow of each, can only be accurately determined by numerous observations at short intervals, say three to five minutes of time apart, from about an hour before to an hour after the actual time of low-water.

"In conclusion, we offer the above observations as solving the problem in question, and as proving the existence of a semi-diurnal lunar tidal wave on Lake Michigan, and consequently on the other great freshwater lakes of North America, whose co-ordinate of altitude is, at its summit, as much as .15 to .25 ($\frac{15}{100}$ to $\frac{25}{100}$ of a foot, United States' measure."

REMARKABLE PHENOMENA.

Prof. Mather, who observed the barometer at Fort Wilkins, Copper Harbor, 47° 30' north lat., during the prevalence of one of these remarkable disturbances which are peculiar to all the Upper Lakes, remarks:—"As a general thing, fluctuations in the barometer accompanied the fluctuations in the level of the water, but sometimes the water-level varied rapidly in the harbor, while no such variation occurred in the barometer at the place of observation. The variation in the level of the water may be caused by varied barometric pressure of the air on the water, either at the place of observation, or at some distant point. A local increased pressure of the atmosphere at the place of observation, would lower the water-level where there is a wide expanse of water; or a diminished pressure, under the same circumstances, would cause the water to rise above its usual level."

In the summer of 1854, according to the Report of Foster and Whitney, made to Congress in 1850, "an extraordinary retrocession of the waters took place at the Saut Ste Marie. The river here is nearly a mile in width, and the depth of water over the sandstone rapids is about three feet. The phenomena occurred at noon; the day was calm but cloudy; the water retired suddenly, leaving the bed of the river bare, except for the distance of about twenty rods where the channel is deepest, and remained so for the space of an hour. Persons went out and caught fish in the pools formed in the rocky cavities. The return of the waters was sudden, and presented a sublime spectacle. They came down like an immense surge, roaring and foaming, and those who had incautiously wandered into the river bed, had barely time to escape being overwhelmed."

Rising and Falling of the Waters of Lake Michigan.

[From the *Chicago Tribune*, May 28, 1861.]

One of those singular oscillations in the Lakes, or "Inland Seas," which have been observed occasionally from the time of the exploration of the Jesuit Fathers, was witnessed yesterday in Lake Michigan. A variety of signs, such as the mirage of the distant shore, unusual depression of the barometer, and a sudden rise of the temperature from a cool, bracing air, to a sultry heat, indicated an unusual commotion in the atmospheric elements. About eleven o'clock A. M., when our attention was first called to the phenomena, the waters of the lake had risen about thirty-one inches above the ordinary level, and in the course of half an hour they again receded. Throughout the whole day they continued to ebb and flow at intervals of fifteen or twenty minutes, and the current between the outer and inner breakwater, near the Illinois Central Railroad House, was so great at times that a row-boat made little or no headway against it. The extreme variation between high and low water was nearly three feet. The wind all day was off shore (from the southwest), the effect of which was to keep down the waters instead of accumulating them at this point. About eight o'clock in the evening it veered suddenly to the northwest, and blew a violent gale, accompanied by vivid electrical displays. This morning (Monday) we hear of telegraphic lines prostrated, of persons killed by lightning, &c., while the lake, although agitated, exhibits none of the pulsations of yesterday.

COMMERCE OF THE LAKES.

Extract from the Annual Report of the Trade and Commerce of Buffalo,

FOR THE YEAR 1862.

"IN presenting to the public our Annual Review of the Trade and Commerce of Buffalo, for the year 1862, it will not be inappropriate to revert to the past to show the rise and progress, the extent and growth of the commerce of those vast 'Inland Seas.'

"When in the year 1679, the Chevalier de La Salle obtained permission of the Seneca Indians to build a vessel at Cayuga Creek, six miles above Niagara Falls, which was launched in 1679, and was the first vessel moved with sails upon the waters of Lake Erie, every portion of the great West was covered with its ancient forests. The echoing axe had never rung through their solitudes, and the battle for mastery was yet undecided between the wild beast and his wild foe the savage hunter. The three guns which were fired when the 'GRIFFIN' was launched, were, probably, the first sounds of gunpowder that ever broke upon the stillness of this vast region. The wondering Senecas heard in them the thunders, and saw the lightnings of heaven. The white man was equally an object of admiration and fear.

"The arts of navigation, at this period, upon this great inland sea, were confined to the bark canoe and the rude paddle with which it was propelled. Never before had the canvas here opened itself to the wind. The voyage of La Salle was an era in the history of this portion of the world. The immense fur trade with the natives at the extremities of these lakes, which was carried on first by the French and afterward by the English, was then almost entirely unknown. It was but the year before the sites of the first trading-houses had been selected. La Salle set sail from the foot of Lake Erie, on the 7th day of August, 1679, with a crew of thirty men, and arrived at Mackinac on the 28th day of that month. The first cargo of furs was put on board the Griffin, and she was ordered by La Salle to return with a crew of six men to Niagara. But a storm was encountered, and the vessel and cargo, valued at fifty to sixty thousand francs, with all on board, was lost. Thus was made the first great sacrifice of life and property to the commerce of Lake Erie.

"Since that period the changes that have been wrought in the country bordering upon and lying beyond these lakes, surpass the dreams of enchantment. Enterprise and energy have penetrated those vast solitudes; the beasts of prey have slunk back into the deep fastnesses of the woods, the native tribes have vanished away like their own majestic forests, and the white man following fast upon their rustling footsteps, has subdued the wilderness to the forms of civilization.

"The country from which the furs were gathered at the trading posts at Niagara,

Detroit, and Mackinac, including a large portion of Ohio, Indiana, Illinois, Michigan, and Wisconsin, now contains a population of 6,926,874. Since the day when La Salle first opened, as it were, to future generations the great highway upon the waters of Lakes Erie and Huron, the progenitors of this mighty multitude have been borne upon its waves by favoring winds; and innumerable little bands gaining the mouth of some fair river, have thence radiated over the wide-spread domain from which their descendants are now pouring down upon the trusting bosom of the lake, the abundant products of an almost inexhaustible fertility.

"Great as has been the change since the country was first explored, it has almost wholly taken place since the year 1800. The population of Ohio in that year was only 45,365; and that was the only State, with the exception of New York and Pennsylvania, of all those bordering upon the great lakes, which contained any considerable settlements, or in which any enumeration of the people was taken. Even Ohio was not then admitted into the Union; and the commercial advantage of Lake Erie were scarcely begun to be developed till twenty-five years afterward. The first vessel bearing the American flag upon Lake Erie was the sloop Detroit, of seventy tons, which was purchased of the Northwest Fur Company, by the General Government, in 1796. She was, however, soon condemned as unseaworthy, and abandoned. Up to the time of the declaration of war in 1812, the whole number of vessels of all descriptions on these lakes, did not exceed twelve, and these were employed either in the fur trade, or in transporting to the West such goods and merchandise as were required for the scattered population that had found their way there. A few vessels were built during the war, but, probably, as many or more were destroyed. And during the three years of its continuance, as all emi-

gration to the West, if any had before existed, must have ceased, there cannot be said to have been any commerce on the lakes.

"In March, 1791, Col. Thomas Proctor visited the Senecas of Buffalo Creek, and from him the first authentic notice of Buffalo is given. He mentions a storehouse kept by an Indian trader named Winne, at Lake Erie.

"In June, 1795, a French nobleman, named La Rochefoucauld Liancourt, visited Buffalo and the neighboring Indian villages. At this place there were then but few houses. He mentions an Inn where he was obliged to sleep on the floor in his clothes.

"In August, 1795, Judge Porter, accompanied by Judah Colt, went to Presque Isle, now Erie, through Buffalo. Judge Porter makes mention 'that one Johnson, the British Indian interpreter, Winne, the trader, and Middaugh, a Dutchman, with his family, lived at Buffalo.' The only road between Buffalo and Avon, in the year 1797, was an Indian trail, and the only house on this trail was one, about one and one-half miles east of the present village of Le Roy, occupied by a Mr. Wilder. As late as 1812 the roads were next to impassable, and to obtain supplies from Albany, trade was carried on by a circuitous route, 'through the Niagara river to Schlosser, thence by portage to Lewiston, thence by water to Oswego and up the Oswego River, through the Oneida Lake and Wood Creek, and across a short portage to the Mohawk River, thence by that river and around the portage of Little Falls to Schenectady—and thence over the arid pine plains to Albany.' The late Judge Townsend and George Coit, Esq., came to Buffalo as traders, in 1811 by this route, bringing about twenty tons of merchandise from Albany at a cost of fifty dollars a ton. At this time there were less than one hundred dwellings here, and the population did not exceed five hundred. The mouth of Buffalo Creek was then obstructed by a sand-bar, frequently preventing the entrance of small vessels, and even frail Indian bark canoes were frequently shut out, and footmen walked across its mouth on dry land. Vessels then received and discharged their cargoes at Bird Island wharf, near Black Rock. To remedy the obstructions in the creek by the sand-bar at its outlet into the lake, it was proposed in the year 1811, to run a pier into the lake, but nothing of moment was done till the spring of 1820, when a subscription was raised, by the then villagers of Buffalo, amounting to $1,361. The late Hon. Samuel Wilkson was the originator and projector of this movement, and temporary improvements were made which carried away the obstructing sandbar. In 1822 the village in its corporate capacity paid John T. Lacy for building a mud-scow for working in the harbor $534. The first corporate notice of the harbor was made in the latter year. Buffalo was incorporated as a village in April, 1813, and as a city on 20th of April, 1832.

"Melish says, 'the population by the last census was 365, and it was computed in 1811 at 500, and is rapidly increasing.' In 1825 the population was 2,412; in 1830, 8,668; in 1835, 15,661; in 1840, 18,213; in 1845, 29,973; in 1850, 42,261; in 1860, 81,129; and at the end of the year 1862 the population is estimated at over 100,000. In 1817 the taxable property of the village was $134,400, and on this valuation an assessment of $400 was made during that year. The valuation of the real and personal property of the city in 1862 is $30,911,014.

"The population and valuation of property, the harbor and harbor improvements, the manufactures and commerce, the canal, railway, and water connections by lake with other portions of the country, the population and productions of the West and Northwest, the large lake, canal, and railway facilities for transportation at the present time, when compared with what they were fifty years ago, 'are marvellous in our eyes,' and if some far-seeing mind, a half century since, had prophesied results of such vast magnitude, he would have been denominated an idle dreamer, and a fit subject for a lunatic asylum.

"The States and Territories bordering on, and tributary to the great lake basin that had fifty years ago but a few thousand population, have now nearly seven millions, which will soon be augmented by the natural increase and by immigration to thirty millions, and Buffalo with its 500 inhabitants in 1811, 81,000 in 1860, will have a population of three or four hundred thousand before the present century shall have passed away. Within the limits of these lake States, where, less than forty years ago, there were neither canals nor railways, there are now 14,484 miles of railway, and 3,345 miles of navigable canals, of which latter about 760 miles are slack-water navigation.

"The whole West and Northwest is now traversed by a net-work of railways, with important canal connections between the different States, where there was a sparsely populated, almost interminable forest or uninhabited prairie. In this march of improvement, making more intimate the social and commercial relations of these widely separated sections of the country, the Empire State has nobly led the way. The far-seeing mind of her honored son, Governor Clinton, projected the Erie Canal, which was completed in 1825, uniting the waters of the Hudson with the lakes. A brighter day then dawned upon the West, the population was rapidly augmented, which was soon succeeded by largely increased agricultural productions that gave new life to commerce. The era of railways was commenced in about the year 1830.

"With these largely increased rail facilities, and the capacity of the New York canals nearly quadrupled, the augmenting facilities do not keep pace with the rapidly augmenting population and largely increased production. Improved channels of communication, both by rail and water, must be made, to enable the producer at the West to get his products more cheaply to market. A country vast in extent, bordering upon the upper Mississippi, the Ohio, Cumberland, Tennessee, Arkansas, Red, and Missouri rivers and their tributaries, and the Red river of the North, traversed by more than twenty thousand miles of navigable waters, will soon be densely peopled; new States to the west of those already admitted will soon knock for admission into the Union; the superabundant products of an almost inexhaustible fertility will be pouring over the lakes and railways, and through the rivers and canals, imparting activity to trade, giving life, strength and vital energy to the largely augmenting commerce of the West. As the star of empire westward wends its way, widening the distance from the great sea-board marts of trade, the prospective wants and increased productions of scores of millions of people will from necessity create cheaper and more expeditious facilities for the transportation of their surplus products to market. There is no country on the face of the globe that has so many natural advantages for a large and extended internal trade as the great West and Northwest.

"The great basin east of the Rocky Mountains is drained by the Mississippi and Missouri Rivers and their tributaries, and their waters find an outlet in the Gulf of Mexico. The great lakes, having an area equal to one twenty-fifth part of the Atlantic Ocean, are drained by the river St. Lawrence, and find an outlet in the Gulf of St. Lawrence. The construction of a few miles of canal makes a navigable connection from the ocean to the great chain of lakes. These natural advantages have been improved to some extent in the United States by the construction of a canal through the State of New York, that now has a prism forty-five feet at the bottom and seventy feet at the top, with seven feet of water, with locks 18 feet 6 inches wide by 100 feet long. There is also a canal one hundred miles long connecting the Illinois river with lake Michigan at Chicago, and slack water navigation connecting Green Bay, Wisconsin, with the Mississippi river. By the construction of a ship canal about three-fourths of a mile in length, from Big Stone Lake to Lake Traver in Minnesota, steamboats from St. Paul could navigate both the Minnesota river and the Red river of the North to Lake Winnepeg, a distance of seven hundred miles. The country traversed by these rivers is surpassingly fertile and capable of sustaining a dense population. Lake Winnepeg is larger than Lake Ontario, and receives the Sas-katch-e-wan river from the West. The Sas-katch-e-wan river is navigable to a point (Edmonton House) near the Rocky Mountains, seven hundred miles west of Lake Winnepeg, and only 150 miles east of the celebrated gold diggings on Frazer river in British Columbia. The digging of that one mile of canal, would, therefore, enable a steamboat at New Orleans to pass into Lake Winnepeg and from thence to Edmonton House, some 5,000 miles. A move has already been made for constructing this short canal. By enlarging the Illinois and Michigan canal and improving the navigation of the Illinois river, and improving and completing the slack water navigation of the Fox river in Wisconsin, connecting Green Bay with the Mississippi river, and still further enlarging the main trunk of the New York canals, steamers could be passed from New York or the Gulf of St. Lawrence, either through the canals of New York or Canada into the great lakes, and from thence to the head waters of the Sas-katch-e-wan, the Missouri, the Yellow Stone rivers, being some 5,000 to 6,000 miles. The cereal product of the States bordering on and tributary to the lakes was 267,295,877 bushels in 1840; 434,-862,661 bushels in 1850, against 679,031,-559 bushels in 1860, and the population of these States has kept pace with their cereal products, being 6,259,345 in 1840; 9,178,003 in 1850, against 13,355,093 in 1860, an increase of nearly fifty per cent. in population and cereal products in each decade. If the same rate per cent. of increase in population and cereal products shall be continued, these States in 1870 will have a population of 20,032,639, with a cereal product of 1,008,557,338 bushels; in 1880, a population of 30,048,958, with a cereal product of 1,512,821,000 bushels; in 1890, a population of 55,073,437, with a cereal product of 2,269,231,510 bushels, and in 1900 a population of 67,610,155, with a cereal product of 3,403,847,265.

"The grain trade of Buffalo for a series of years, given in this report, has already reached upwards of 72,000,000 bushels for the year 1862. If a crop of 680,000,000 of bushels of cereal products, gives Buffalo 72,000,000 of bushels of that crop; in the year 1900, with a crop of 3,403,-847,265 bushels of cereal products, the grain trade of Buffalo will be upward of 360,000,000 of bushels. The calculations of the forty years of the future are based on the actual results of the last thirty years. The year 1870 will give to Buffalo a grain trade of upwards of 107,000,000 of bushels, and if there shall be a proportionate increase in the grain trade of Oswego, the present capacity of the New York canals will be entirely inadequate to pass through them this large amount of grain in addition to the large increase in the tonnage of other commodities, saying nothing of the capacity that will be required for the augmented business in 1880, 1890, and 1900."

The First Steamboat on Lake Erie.

The Detroit *Tribune* furnishes some interesting extracts on this subject, taken from the files of the Detroit *Gazette*, of 1818. We select the following description of the reception of this monster of the great deep by the "*Wolverines*" of that day.

"AUGUST 26, 1818:—Yesterday, between the hours of 10 and 11 A. M., the elegant steamboat *Walk-in-the-Water*, Capt. J. Fish, arrived.—As she passed the public wharf, and that owned by Mr. J. S. Roby, she was cheered by hundreds of the inhabitants, who had collected to witness this (in these waters) truly novel and grand spectacle. She came to at Wing's wharf. She left Buffalo at half-past 1 o'clock on the 23d, and arrived off Dunkirk at 35 minutes past 6 on the same day. On the next morning she arrived at Erie, Capt. Fish having reduced her steam during the night, in order not to pass that place, where she took in a supply of wood. At half-past 7 P. M. she left Erie, *and came to at Cleveland* at 11 o'clock. On Friday, at 20 minutes past 6 o'clock, P. M., sailed and arrived off Sandusky Bay at 1 o'clock on Wednesday; lay at anchor during the night, and then proceeded to Venice to wood; left Venice at 3 P. M., and arrived at the mouth of the Detroit River, where she anchored during the night—the whole time employed in sailing, in this first voyage from Buffalo to this, being about 44 hours and 10 minutes; the wind ahead during nearly the whole passage. Not the slightest accident happened during the voyage, and all our machinery worked admirably.

"Nothing could exceed the surprise of the sons of the forest on seeing the *Walk-in-the-Water* moving majestically and rapidly against a strong current, without the assistance of sails or oars. They lined the banks near Malden, and expressed their astonishment by repeated shouts of ' Tai-yoh nichee.' A report had been circulated among them, that a 'big canoe' would soon come from the noisy waters, which, by order of the great father of the Cho-mo-ko-mons, would be drawn through the lakes and rivers by sturgeon! Of the truth of the report they are now perfectly satisfied. The cabins of this boat are fitted up in a neat, convenient, and elegant style; and the manner in which she is found, does honor to the proprietors and to her commander. A passage between this place and Buffalo is now not merely tolerable, but truly pleasant. To-day she will make a trip to Lake St. Clair, with a large party of ladies and gentlemen. She will leave for Buffalo to-morrow, and may be expected to visit us again next week."

LAKE COMMERCE.
Commerce of Buffalo.—1862.

The Collector of Customs for Buffalo has furnished the following statement, showing the arrivals and clearances of American and Foreign vessels to and from Canadian ports; also, the arrivals and clearances of American vessels to and from American ports; the tonnage of same; and the number of men employed:

American vessels:	No.	Tonnage.	Crew.
Entered, 1st quarter.....	320	354,000	4,160
" 2d "	366	294,241	3,344
" 3d "	308	212,805	2,587
" 4th "	342	302,929	3,295
Foreign vessels:			
Entered, 1st quarter......	....	....	....
" 2d "	219	20,836	1,049
" 3d "	321	25,632	1,560
" 4th "	145	13,705	719
American vessels:			
Cleared, 1st quarter	323	384,185	4,168
" 2d "	399	294,755	3,461
" 3d "	325	213,365	2,685
" 4th "	334	294,526	3,197
Foreign vessels:			
Cleared, 1st quarter	....	....	....
" 2d "	205	19,572	987
" 3d "	309	28,345	1,536
" 4th "	138	13,273	687
Coasting vessels:			
Entered, 1st quarter......	....	....	....
" 2d "	1,802	601,673	19,364
" 3d "	2,772	920,979	29,276
" 4th "	1,611	576,354	17,792
Cleared, 1st "	4	2,129	76
" 2d "	1,989	657,183	21,082
" 3d "	2,733	907,357	28,825
" 4th "	1,480	518,318	16,250

SUMMARY FOR THE YEAR.

	No.	Tonnage.	Crew.
American vessels entered.	1,331	1,193,975	13,386
Foreign vessels entered...	685	63,173	3,328
Coasting vessels entered..	6,185	2,099,006	66,432
Total entered for the year.	8,201	3,356,154	83,146
American vessels cleared.	1,381	1,186,831	13,511
Foreign vessels cleared...	652	61,195	3,210
Coasting vessels cleared..	6,156	2,085,011	66,233
Total cleared for the year.	8,189	3,333,037	82,954
Grand Total, 1862........	16,390	6,689,191	166,183
" 1861......	18,866	5,963,996	144,173
" 1860......	11,527	4,710,175	120,497
" 1859......	10,521	5,592,626	118,109
" 1858........	8,818	3,329,246	86,851
" 1857........	7,551	3,221,806	132,189
" 1856........	8,128	3,018,589	112,051
" 1855........	9,211	3,360,238	111,515
" 1854........	8,912	3,990,284	120,838
" 1853........	8,295	3,252,973	128,112
" 1852........	9,441	3,092,247	127,491

United States and Canadian Tonnage.

The following statements from the report of the Secretary of the Board of Lake Underwriters for 1862, will show the tonnage, value, and class of vessels navigating the Northwestern Lakes in 1861 and 1862, viz.:

Comparative statement of the tonnage of the Northwestern Lakes, and the river St. Lawrence, on the 1st day of January, 1862 and 1863:

1862.

Class of vessels.	No.	Tonnage.	Value.
Steamers...........	147	64.669	$2,668,900
Propellers..........	203	60,951	2,814,900
Barks......	62	25,113	621,440
Brigs...............	86	25,871	501,100
Schooners..........	939	204,900	5,248,900
Sloops	15	2,900	11,850
Barges..............	....	....	......
Totals........	1,502	358,309	$11,862,450

1863.

Class of vessels.	No.	Tonnage.	Value.
Steamers....	143	58,522	$2,190,300
Propellers..........	254	70,253	3,573,800
Barks	74	83,203	982,900
Brigs	85	21,831	526,200
Schooners..........	1,069	227,831	5,955,350
Sloops	16	667	12,770
Barges.............	2	8,719	17,000
Totals	1,643	413,026	$13,257,020

Increase in number of vessels	141
Increase in tons....................	29,717
Increase in value..................	1,394,570

TABLE,

EXHIBITING THE TONNAGE *of the several Lake Districts in the United States, on the 30th June,* 1861.

Districts.	State.	Lakes, &c.	Total Tonnage.
Burlington	Vermont	Champlain..........	$7,774 19
Champlain....	New York	"	1,791 71
Oswegatchie.......	"	St. Lawrence River..	7,332 53
Cape Vincent.......	"	" ..	5,223 70
Sackett's Harbor....	"	Ontario	888 55
Oswego.............	"	"	55,552 41
Genesee	"	"	2,981 84
Niagara.............	"	"	774 48
Buffalo.............	"	Erie	108,224 00
Dunkirk	"	"	4,274 26
Presque Isle (Erie)..	Pennsylvania	"	7,369 09
Cuyahoga (Cleveland)	Ohio...............	"	82,518 87
Sandusky	"	"	15,850 24
Toledo.............	"	"	5,468 70
Detroit	Michigan...........	Detroit River	66,887 89
Michilimackinac	"	Huron	4,747 59
Chicago............	Illinois............	Michigan...........	85,743 66
Milwaukee	Wisconsin..........	"	27,048 19
	Minnesota	Superior	
Total Tonnage................................			$500,456 90

FLOUR AND GRAIN,

FLOUR AND GRAIN TRADE OF DIFFERENT CITIES COMPARED.

The importance of Buffalo as a grain receiving port, will be shown by the following comparative statements of the grain trade of Lake cities with several of the grain ports of Europe:

BUFFALO.

	1860.	1861.	1862.
Flour, bbls	1,122,335	2,159,591	2,846,022
Wheat, bush....	18,502,649	27,105,219	30,435,831
Corn, bush......	11,386,217	21,024,657	24,258,627
Oats, bush......	1,209,594	1,797,905	2,624,932
Barley, bush....	262,158	313,757	423,124
Rye, bush	80,822	387,764	791,564
Total grain	41,441,440	50,597,302	58,564,078

DUNKIRK.

	1860.	1861.	1862.
Flour, bbls......	542,765	736,529	1,095,864
Wheat, bush....	500,888	604,561	112,061
Corn, bush	644,081	230,400	149,654
Oats, bush .... }		3,950	
Barley, bush.. }	8,843	....	10,173
Rye, bush .... }		3,225	
Total grain	1,153,812	842,136	271,888

CHICAGO.

	1860.	1861.	1862.
Flour, bbls......	713,348	1,479,284	1,755,258
Wheat, bush....	14,427,083	17,385,002	13,187,583
Corn, bush	15,262,894	26,369,989	31,145,721
Oats, bush......	2,198,889	2,067,018	3,782,422
Barley, bush....	617,619	457,589	800,476
Rye, bush	318,976	490,989	976,752
Total grain	32,824,961	46,770,587	49,842,904

OGDENSBURGH.

	1860.	1861.	1862.
Flour, bbls......	248,200	411,888	576,394
Wheat, bush....	665,022	677,386	689,930
Corn, bush.	867,014	1,119,594	1,120,176
Oats, bush......	28,242	2,365	3,336
Barley, bush....	7,105	15,151	15,529
Rye, bush	3,050	3,888	....
Total grain	1,470,433	1,818,384	1,828,974

MILWAUKEE.

	1860.	1861.	1862.
Flour, bbls	805,208	492,259	503,957
Wheat, bush....	9,108,458	15,930,706	14,253,853
Corn, bush	126,404	114,931	265,128
Oats, bush	178,963	131,256	289,380
Barley, bush....	109,795	66,991	141,359
Rye, bush......	52,382	73,448	159,512
Total grain	10,576,002	16,817,332	15,109,232

TOLEDO.

	1860.	1861.	1862.
Flour, bbls......	807,768	1,406,676	1,585,325
Wheat, bush....	5,341,190	6,277,407	9,827,629
Corn, bush	5,386,951	5,812,038	3,813,709
Oats, bush....	129,689	41,428	234,759
Barley, bush....	115,992	12,064	63,038
Rye, bush	37,787	31,193	44,368
Total grain ....:	11,011,609	11,674,130	13,983,593

OSWEGO.

	1860.	1861.	1862.
Flour, bbls	121,399	119,056	235,382
Wheat, bush....	9,651,564	10,121,446	10,982,132
Corn, bush	5,019,400	4,642.262	4,528,962
Oats, bush......	858,416	116,884	187,284
Barley, bush....	1,326,915	1,173,551	1,050,864
Rye, bush	244,311	381,687	130,175
Total grain	16,630,606	16,435,330	16,878,917

CAPE VINCENT.

	1860.	1861.	1862.
Flour, bbls......	28,940	65,407	48,576
Wheat, bush....	203,879	276,610	816,403
Corn, bush	73,800	124,411	249,369
Oats, bush	27,299	2,904	1,030
Barley, bush ...	90,614	53,877	31,265
Rye, bush	20,616	23,365	762
Total grain ...	415,707	481,257	598,829

DETROIT.

	1860.	1861.	1862.
Flour, bbls	862,175	1,321,140	1,543,876
Wheat, bush....	1,809,523	2,505,111	3,058,242
Corn, bush	608,698	1,036,506	583,861
Oats, bush......	319,598	388,956	407,247
Barley, bush....	124,882	59,734	165,200
Rye, bush	80,843	16,981	18,807
Total grain	2,923,544	4,007,318	4,283,357

Summary of Receipts, 1862.

	Flour, bbls.	Grain, bush.
Buffalo	2,846,022	58,564,078
Chicago	1,755,258	49,842,904
Milwaukee	503,957	15,109,232
Oswego	235,382	16,878,917
Detroit..................	1,543,876	4,283,357
Dunkirk.................	1,095,864	271,888
Ogdensburgh	576,394	1,828,974
Toledo	1,586,325	13,983,593
Cape Vincent	48,576	598,829

The grain trade of the great West and Northwest is yet in its infancy. Every year the population is augmented by emigration from the Eastern and Middle States as well as from Europe. The strong arms of freemen are bringing under cultivation the broad prairies upon which the industrious and enterprising settlers scatter broad-cast the seed, to be returned to them again in fields of waving grain, from which will be reaped a bountiful harvest.

Imports of Breadstuffs into Great Britain.

The following from the London *Times* will show the imports of breadstuffs into Great Britain for the years 1860, 1861, and 1862, ending December 31st in each year, viz.:

	1860.	1861.	1862.
Wheat, Flour, Meal, cwt......	5,139,188	6,331,375	7,314,317
Wheat, qrs. of 8 bush.........	5,903,175	6,966,844	9,542,362
Corn. " "	1,895,594	3,106,595	2,751,265
Oats, " "	2,308,380	1,875,574	1,622,919
Barley, " "	2,122,016	1,407,501	1,863,683
Rye, " "	96,838	54,142	1,694
Peas, " "	317,548	402,933	230,132
Buckwheat, " "	....	714	5,143
Beans, " "	440,860	564,477	479,758
Total Grain, qrs. 8 bush	13,044,471	14,377,780	16,496,956

The *Times* says: "The accounts have now been made up of the quantities of grain and flour imported into Great Britain on the last year and preceding years since the introduction of free trade, and the result is remarkable, showing an extraordinary increase during the past year. Taking wheat and flour alone, we find that the lowest year was 1835, when the quantity of wheat imported into Great Britain was only 46,530 quarters, and of flour 84,684 cwt., while in 1862, no less than 9,541,362 quarters of wheat, and 7,314,317 cwt. of flour were imported into the various ports of the country."

The increase in 1862 over 1861 is 982,-942 cwt. of flour, and 2,575,518 quarters of wheat. There is a decrease on Indian corn of 355,330 quarters.

FISHERIES—FISH.

In the Sandusky bay, in the Maumee bay and Maumee river, in the Monroe bay, in the Detroit river, in the St. Clair river and rapids, in Lake Huron, from Huron to Point aux Barques, in the Au Sable river, in Thunder bay above Au Sable river, including Sugar Island, in Saginaw bay and river, in Tawas bay, between Thunder bay and Mackinac, including Hammond's bay, in and about Mackinac at Beaver Island and its surroundings, between the De Tour and the Sault, along the Eastern shore of Lake Michigan, in Green bay, in Wisconsin and Michigan, at Presque Isle, Pa., in Lake Superior's numerous bays and inlets, are found the principal fishing

3

grounds of the lakes; and the annual catch ranges from 60 to 100 thousand barrels, valued at four to six hundred thousand dollars. The lake fisheries are only second to the cod fisheries off the Atlantic coast, from Cape Cod Bay to Cape Breton, and are a source of very considerable wealth.

The stock here will, probably, not exceed twenty-five hundred packages:

LAKE IMPORTS OF FISH.

Years.	Bbls.	Years.	Bbls.
1854	11,752	1859	13,391
1855	7,241	1860	26,655
1856	6,250	1861	8,313
1857	5,290	1862	8,647
1855	4,208		

LUMBER AND STAVES.

The Lumber and Stave trade constitutes a very large portion of the freight carried on the lakes and canals, and is only second to grain. The larger portion of the Eastward movement usually takes place in midsummer, when low rates of transportation rule. The principal sources of supply are the States of Ohio, Indiana, Michigan, Canada West, and Pennsylvania, of which more than fifty per cent. is from Michigan alone. In the northern peninsula of that State, in and around Saginaw, at Port Huron, on St. Clair river, are the largest and finest lumber districts in the West and Northwest.

The supply of staves is derived from Ohio, Indiana, Michigan, Wisconsin, and Canada West, of which more than eighty per cent. of the receipts at Buffalo come from the States first mentioned.

The following will show the imports at Buffalo of staves and lumber from 1846 to 1862, inclusive, and the canal exports from 1849 to 1862, inclusive:

LAKE IMPORTS.

Years.	Staves, No.	Lumber, feet
1846	10,762,500	84,586,000
1847	8,800,000	18,313,000
1848	8,091,000	21,425,000
1849	14,183,602	33,935,768
1850	18,652,890	53,076,000
1851	10,696,006	63,006,000
1852	12,998,614	72,337,225
1853	9,215,240	89,294,000
1854	15,464,554	67,407,008
1855	16,421,568	72,026,651
1856	18,556,089	60,584,812
1857	23,024,213	68,283,319
1858	15,119,019	67,059,178
1859	23,277,028	111,072,446
1860	22,307,839	111,094,496
1861	25,228,973	58,082,713
1862	30,410,252	125,289,971

COPPER—LAKE SUPERIOR.

The Copper Mines of Lake Superior were first brought into public notice in 1845, when speculation was rife in all that spur of the Porcupine Mountains on the south shore of Superior, extending far into the lake, having for its base a line drawn across L'Anse Bay to Ontonagon. This was then the Northern El Dorado. In this year operations were commenced at the Minnesota mine, which is about fifteen miles back of Ontonagon. The first large mass of native copper, weighing about seven tons, was found in a pit dug by the original lords of the soil.

It is now only fourteen years since this mine was opened. At that time the rapids in the Sault St. Marie prevented the passage of vessels from the lower lakes, and the adventurers that sought out this new El Dorado, had many obstacles to overcome. The country was then covered by a vast wilderness, without inhabitants, excepting a few Indians.

All supplies were brought from the lower lakes, and then had to be passed over the Portage at St. Mary, and thence carried in frail vessels coasting to the westward, hundreds of miles to the copper regions, and then carried on the back of man and beast to the supposed places of the copper deposit. Every stroke of the pick was made at a cost ten-fold more than in populated districts, every disaster delayed operations for weeks and even months.

The opening of the St. Mary canal, in 1856, has produced a wondrous change in all this wilderness region. The only settlements on the south shore of the lake, at the present time, are Marquette, Portage Lake, Ontonagon, Copper Harbor, Eagle Harbor, Eagle River, and the adjacent mines—all else is a vast wilderness, without sign of human habitation.

The Copper region is divided into the three districts of Ontonagon, Keweenaw Point, and Portage Lake. Since 1845, 120 Copper Mining Companies have been organized under the General Law of Michigan; more than six millions of dollars have been expended in explorations and mining improvements. The Minnesota and Cliff mines have declared and paid over two millions of dollars in dividends since the organization and working of these companies.

Until 1860 all the Copper of the Lake Superior mines was smelted at Detroit, Cleveland, and Boston. Since which time a Boston company have erected smelting works at Portage, Lake Superior, while some Copper Ore has been shipped to Liverpool to be smelted there.

There is an annual product of Copper of about 2,500 to 5,000 tons at the Wellington mines, Lake Huron, which are worked by a Canadian company.

BUSINESS ON LAKE SUPERIOR IN 1862.

The annual report of the Superintendent of the Sault St. Mary Canal, to the Governor of Michigan, says that during the last year there passed through the ship canal $12,000,000 worth of copper and iron, and general merchandise to the value of $10,000,000. The number of vessels, sail and steam, that passed through the canal was 838, and the aggregate tonnage was 349,612 tons. In 1861 the number of vessels was 527, and the tonnage 276,637 tons. The tolls collected on the canal are six cents per ton, making an income of $21,676 72. The trade last year may be divided as follows: Iron, pig and ore, 150,000 tons; copper, 9,300 tons; general merchandise, 8,000 tons.

The following, showing the shipments and value of Copper shipped from Lake Superior from 1845 to 1862, inclusive, will indicate the growth and importance of the Copper mining interest of Lake Superior:

AGGREGATE SHIPMENTS OF COPPER FROM LAKE SUPERIOR FROM 1845 TO 1862.

Shipments in	Tons. Lbs.	Value.
1845....	.1300	$200
1846....	29.	2,619
1847....	239.	107,550
1848....	516.	206,400
1849....	750.	301,200
1850....	640.	266,000
1851....	872.	348,800
1852....	887.	800,450
1853....	1,472.	588,200
1854....	2,300.	865,000
1855....	3,196.	1,437,000
1856....	5,726.	2,400,100
1857....	5,759.	2,015,650
1858....	5,996.	1,610,000
1859....	6,041.	1,932,000
1860....	8,614.	2,520,000
1861....	10,337.	8,150,000
1862....	*10,000.	4,000,000

SHIPMENTS OF THE COPPER DISTRICTS— FOUR YEARS.

	1859.	1860.	1861.	1862.
Keweenaw Dist.	1,910.3	1,910.8	2,151.9	2,726.6*
Portage Lake...	1,583.1	3,064.6	4,705.6	4,288.9*
Ontonagon.....	2,597.6	3,610.7	8,476.7	2,706.1
Carp Lake......		20.5		7.1
Sundry Mines..		7.6		

The Copper product of Lake Superior, although small when compared with the product in Great Britain, has, since 1845, when the mines were first worked, grown into a trade of large proportions, the aggregate value of product from 1845 to 1862, inclusive, being about $21,911,300.

* Estimated.

IRON ORE AND IRON—LAKE SUPERIOR.

The connecting of the waters of Lake Superior with the waters of Lake Huron, by the construction of a ship canal three-fourths of a mile in length, around the rapids in the Sault St. Mary river, with prism and locks of sufficient capacity for passing the largest class of vessels navigating the lakes, completed very soon after the discovery and working of the iron mines, has opened an already extensive commerce in iron ore, and pig-iron manufactured near the mines, which are sixteen to eighteen miles from Marquette. These mines are about seven hundred feet above the level of the lake, and are connected with Marquette by a railroad.

To show the rise and progression of this trade, we give below an interesting statement from the Marquette *Journal* of January 16, 1863:

THE IRON PRODUCT OF THE LAKE SUPERIOR —SHIPMENTS OF IRON ORE.

Year.	Jackson Iron Co.	Cleveland Iron Co.	Lake Sup'r Iron Co.	Total Gross Tons.
1855....		1,447	....	1,447
1856.... 4,497		7,100	....	11,597
1857....13,912		12,272	....	26,184
1858....11,104		19,931	....	31,035
1859....10,662		30,344	24,668	65,679
1860....41,286		42,696	33,016	116,998
1861....12,919		7,311	25,200	45,430
1862....42,767		35,244	87,710	115,721

Total amount shipped to date.......414,091

PRODUCTS OF PIG-IRON.

	Pioneer Iron Co.	Collins Iron Co.	Forest Iron Co.	Northern Iron Co.*	Value lb. ton.
1858.... 1,627		....	....	....	$25 00
1859.... 4,683		2,575	....	....	25 00
1860.... 3,560		1,950	....	150	25 00
1861.... 2,550		2,060	2,430	900	23 00
1862.... 1,438		2,207	2,802	2,143	85 00

* Estimated.

Date.	Tons Ore, gross.	Tons Pig, gross.	Total Value.
1855............	1,447	....	$14,470
1856............	11,597	....	92,776
1857............	26,184	....	209,472
1858............	31,035	1,627	249,269
1859............	65,679	7,258	575,521
1860............	116,998	5,660	736,490
1861............	45,430	7,970	410,460
1862............	115,721	8,590	984,976

It will be seen from this statement that the shipments of iron in 1862 were 115,721 tons against 1,447 tons in 1855; and 8,590 tons of pig-iron in 1862, against 1,627 tons in 1858.

Marquette is the only point on Lake Superior where iron mines have been opened, although there are iron deposits in the mountains back of L'Anse. About eighteen miles from Marquette are the *Iron mountains* named the Lake Superior, the Jackson, the Burt, the Collins, the Barlow, and the Cleveland, while eight miles further back are the St. Clair and Ely mountains. Only three of these are at present worked, the Jackson, the Cleveland, and the Lake Superior, but these alone contain enough iron to supply the world for many generations. Still further back from the lake rise mountains to eight hundred feet high, covering many hundreds of acres, which, it is believed from explorations already made, are solid iron ore. There are now in operation at Marquette three iron mining companies, and two blast furnaces, the Pioneer and Collins, for making charcoal pig-iron. The Collins has one stack, and can turn out about eleven tons of pig-iron daily; the Pioneer has two stacks, with a capacity for the manufacture of about twenty tons daily. About three miles to the south of Marquette, at the mouth of the Chocolate River, the Northern Iron Company have quite recently built a large bituminous coal furnace.

The quality of the Lake Superior iron is conceded, by all who have given it a trial, to be superior to any iron in the world, as is shown by the following analysis by Prof. Johnson, giving the strength per square inch in pounds:

Salisbury, Conn., Iron	58,000
Swedish (best)	58,184
English Cable	76,105
Centre County, Pa	59,400
Essex County, New York	59,962
Lancaster County, Penn	76,069
Common English and American	80,000
Lake Superior	89,582

Large quantities of iron ore are taken from Marquette, Mich., to Detroit, Cleveland, Erie, and Huron, while several thousand tons have annually been brought to Buffalo for smelting, or to pass through the Erie, Seneca, and Chemung Canals to amalgamate with the iron ores of Pennsylvania, for the manufacture of pig-iron. Two extensive establishments have been already erected in Buffalo, which can smelt annually from fifteen to twenty thousand tons of ore into pig iron. There will soon be another smelting furnace erected here, which will, probably, be in operation before the end of the present year. In addition to these, a large rolling mill for rolling railroad and bar iron has just been completed, and is now in full operation.

There is in the Lake Superior iron district enough iron ore to supply the world with iron. A new era in ship and boat building is near at hand. Iron vessels will take the place of wood, when a large amount of iron plates and beams will be required to construct iron ships and steamers. There is a network of railways centring here, which will require a very large amount of railroad iron to replace that now in use.

The opening of the iron trade of Lake Superior will, in the future, have an important bearing upon the trade of the New York canals. The ores of Clinton, Oneida, are required to mix with the ore of Lake Superior. The East will soon obtain its supply of pig, bar, and railroad iron from the West.

With better and cheaper facilities for the transportation on the Erie canal, the tonnage will assume a magnitude and importance commensurate with the demand for this all-important mineral product.

LAKE IMPORTS.

IRON ORE		PIG IRON	
	Tons		Tons
1859	555	1859	1,404
1860	3,728	1860	8,795
1861	3,563	1861	1,563
1862	10,034	1862	8,465

SALT.

In the year 1860, there were manufactured in the eight following named States 12,190,953 bushels of salt, of which there were produced 30,900 in Massachusetts, 7,521,335 bushels in New York; 604,300 bushels in Pennsylvania, 1,744,240 bushels in Ohio, 2,056,513 bushels in Virginia, 69,665 bushels in Kentucky, 120,000 bushels in Texas, and 44,000 bushels in California, valued at $2,265,302.

The discovery of salines in Michigan, at Saginaw and vicinity, in 1859, will soon add largely to the salt product. The East Saginaw Salt Company, with a capital of fifty thousand dollars, was organized in April, 1859, and operations were commenced about the first of May in that year, when they commenced sinking a well, which well was completed to the depth of 670 feet about the first of February, 1860. Immediately after the completion of this well, small quantities of salt were produced in a temporary arrangement, with three or four ordinary kettles, but the manufacture of salt as a business, was not in full operation till July, 1860, when a block of fifty kettles was completed. This company has six blocks of kettles in operation, besides 500 solar vats. There are now in operation 53 blocks of kettles, in addition to which forty-seven firms and companies have been organized, and many of them have wells completed, or nearly completed, and will soon have many additional works.

The works already in operation are capable of producing 870,525 bbls. annually; and before the first of September next, the works being constructed will augment the capacity to one and a quarter millions of barrels annually.

The manufacture of salt from the salines of the Onondaga Salt Springs was commenced as early as the year 1797, which is the date of the first leases of lots, and during that year 25,474 bushels of salt were manufactured. Passing over a period of ten years, to 1807, in which year there were manufactured 165,448 bushels; in 1817, 448,665 bushels; in 1827, 983,410 bushels; in 1837, 2,161,287 bushels; in 1847, 3,951,351 bushels; in 1851, 4,614,117 bushels; in 1861, 7,300,000 bushels; in 1862, 9,016,-013 bushels. These salines have supplied the Eastern, Middle, Western and North-western States for many years, and in a little more than half a century the product has been augmented from 25,474 bushels in 1797, to 9,016,013 bushels in 1862; making the total product since 1797 upwards of 145,000,000 of bushels. The United States government duties on foreign salt, and the adoption of a higher rate of toll than on domestic salt, have protected the manufacture to such an extent, that foreign salt (until 1862, when the rate of toll on it was reduced) was almost entirely excluded from finding a consumptive demand. The salt manufacture in New York has given employment to a large number of our people, and has largely augmented both the tonnage and revenue of the New York canals. The outlets for our domestic salt, going to Western States and Canada, is by the way of Buffalo and Oswego, which latter place has always received, since the completion of the Oswego canal, much the larger share for the annual supply of the West.

The following will show the imports of salt by canal at Buffalo and Oswego for the years indicated:

IMPORTS OF DOMESTIC SALT.

Years.	Buffalo, lbs.	Oswego, lbs.
1849	89,952,000	113,184,000
1850	25,612,000	69,090,000
1851	30,084,000	113,742,000
1852	44,316,000	102,164,000
1853	59,327,474	....
1854	67,557,072	168,410,000
1855	109,325,811	148,110,000
1856	60,913,378	193,684,000
1857	52,228,989	142,967,755
1858	77,001,105	243,709,816
1859	112,621,028	190,262,431
1860	92,949,269	159,527,670
1861	159,191,278	173,193,476
1862	177,620,485	228,698,389

What effect upon the salt manufacture and the canal commerce of this State, the discovery and working of the salines of the Saginaw Valley in Michigan will have, remains to be seen. The favorable accounts of the strength of these newly discovered salines, and the progress already made in the manufacture of salt, induce the belief that Michigan will soon be a strong competitor with New York in the markets of the West for this almost universally used commodity.

PETROLEUM.

During the last two years Petroleum has assumed an importance in the economy and material interests of the country that is as yet but partially developed. The lately discovered use of this crude commodity, that is found deposited in large reservoirs from five to seven hundred feet beneath the surface of the earth, which, when tapped, flows upward to the surface in almost inexhaustible supply, is producing a revolution in the economies of the people's light. In Canada West, in the vicinity of Sarnia, and on Oil Creek in Pennsylvania, these supply reservoirs have already been largely developed, and new discoveries will, probably, show a large extent of territory where this crude commodity can be obtained.

In the Pennsylvania oil regions there are seventy-five flowing wells, sixty-two wells that formerly flowed and were pumped, besides three hundred and fifty-eight wells sunk and commenced, costing, on an average, one thousand dollars each, equal to $495,000. In the vicinity of these wells are twenty-five refineries for refining the crude Petroleum, which, with the machinery and buildings, have been erected at a cost of about $500,000. The daily product of the Pennsylvania wells is about 4,400 barrels.

Some of the flowing wells of this oil district have produced as high as from two to seven hundred barrels of crude oil daily.

There are in Canada West thirty-seven firms engaged in refining the crude Petroleum produced near Sarnia, with a weekly product of about 2,200 bbls. of refined oil from seventy-three stills, which will give an annual product of refined oil, if run to their full capacity, of 32,120,000 gallons. The annual product of the Pennsylvania oil regions, at about 4,400 barrels daily, will be about 58,400,000 gallons.

FOREIGN EXPORT OF PETROLEUM IN 1862.

From	Gallons.	Value.
New York	6,783,568	$2,037,413
Philadelphia	2,607,308	527,575
Boston	891,615	457,859
Canada	1,279,000	255,800
Totals	11,561,351	$3,250,647

The exports from the port of New York in 1861, were only 1,112,250 gallons. There were carried Eastward from the Pennsylvania oil regions in 1862, over the Pennsylvania Central railroad, from

Pittsburg, destined for Philadelphia and Baltimore, 73,658 tons, being equal to about 442,000 bbls. of oil.

In every considerable town in the Middle and Western States, there are oil refineries for manufacturing the crude Petroleum.

This discovery has left Coal Oil, Camphene, Lard and Whale Oils at a discount. A cheap light of great brilliancy has been obtained for the million, and its discovery and introduction into general use will, in a very considerable measure, affect prices of other commodities previously used for illuminating purposes, and will give a respite to the Whale, and bids fair to rival even coal gas in cheapness and illuminating power.

The refining of crude Petroleum has already obtained a very considerable magnitude in Buffalo. Some eleven refineries have been erected in this city within the last two years, in which $125,000 to $130,000 in capital has been invested, giving employment to upwards of a hundred persons, having a capacity to refine 75,000 to 80,000 barrels of the crude oil annually.

In all the busy marts of trade, in every considerable town in the Eastern, and Middle, and Western States, the odorous Petroleum meets the olfactories of the passers-by.

What the results of this important discovery will be during the next ten years remains to be seen. If the supply is inexhaustible, and the developments of the past two years are any criterion for the future, the half has not been told of its importance to the world. It has already become an important article in foreign and domestic commerce. The foreign export is already counted by millions, and in its distribution supplies nearly all the important countries of Europe, Australia, and other places in the East Indies, California, some of the South American States, and the Islands of the Pacific.

CANAL COMMERCE.

The construction of the *Erie Canal* was commenced in the year 1817, and the waters of Lake Erie were united with the waters of the Hudson river on the 26th of October, 1826. The first revenue from the Oswego canal was received in 1828. This work was at first suggested as early as 1816, by a memorial from the city of New York to the Legislature. As early as 1724, Cadwallader Colden, then Surveyor-General of the province of New York, described the route as practical to Lakes Champlain and Ontario.

The Erie Canal is one of the largest and most important canals in the world. Notwithstanding the contracted scale of the first structure, the predictions of its projector, DeWitt Clinton, have been more than verified.

This work was urged by Gov. DeWitt Clinton in 1791, and in 1792 by General Schuyler; by Surveyor-General DeWitt in 1808, and at every meeting of the Legislature till 1817, when the work was commenced. In 1810 Governeur Morris, Stephen Van Rensselaer, De Witt Clinton, Peter B. Porter, and others, were, by joint resolution of the Legislature, appointed as Commissioners for exploring the route from the Hudson river to Lakes Ontario and Erie.

In 1812 these Commissioners, in their report to the Legislature, estimated that, in 1832, there would be 250,000 tons brought down the canals, which estimate fell very considerably short of the amount carried on the canals during that year.

This work is 352 miles in length, from Buffalo to Albany, and 345 miles from Buffalo to Troy. The rise and fall from Lake Erie to the Hudson is 692 feet. The prism was originally 40 feet wide at the surface, and 28 feet at the bottom, and four feet deep, with locks of sufficient size to pass boats drawing 3½ feet of water, 14 feet beam by 80 feet long.

The original cost of the first structure was $9,027,456.

In 1835 the Legislature passed an act providing for its enlargement. Under this and other acts, the prism of the canal has been increased so as to be 70 feet on the surface, 42 feet on the bottom, and 7 feet in depth, with locks 110 feet long by 18 feet wide, passing boats 96½ feet long by 17½ feet beam, drawing 5 feet 10 inches to 6 feet of water.

There are now 71 locks on the Erie canal between Buffalo and Albany, and 18 locks on the Oswego canal between Syracuse and Oswego.

The enlargement is now completed after the plan adopted in 1835, with some modifications, at an additional cost of about fifty millions of dollars.

Connecting with the main trunk of the Erie canal are the Champlain, 64 miles long; the Chenango, 98 miles; the Black River, 103 miles; the Chemung canal; the Genesee Valley canal, besides several other connecting links, making, with the Erie and Oswego canals, a total of 1,028 miles of canal, including 100 miles of slack water navigation connected by canals, within and belonging to the State of New York.

COST AND REVENUES OF THE NEW YORK CANALS.

The State Auditor, in his report to the Legislature in 1863, makes a balance sheet, from which the following summary of the revenue from all sources, from 1817 to and including September 30th, 1862, and the disbursements for all purposes for the same period, has been made, viz.:

RECEIPTS.

Gross canal tolls.....		$81,068,959 45
Direct taxes on the people		8,806,881 94
Indirect taxes, viz:		
Auc. dut's $3,592,080 05		
Salt duties 2,055,455 06		
Tax on st'mbout passeng. 73,500 99		
		5,721,007 10
*General Fund for Deficiencies..........	1,386,496 83	
Sales canal lands....	320,515 15	
Interest on investments and deposits.	8,723,417 12	
Premiums on loans..	2,294,504 23	
Rents, surplus wt's..	89,421 74	
Elmira and Chemung Canal Feeder......	290,097 66	
Miscellaneous sources	1,569,709 88	$104,791,104 65

DISBURSEMENTS.

Prems. on purchase of stocks, commis., &c.	$366,799 79	
Interest on loans.....	81,821,403 12	
Contractors, Collectors, Weighmasters, &c..	22,561,813 93	
Commissioners for construction.......	61,213,596 85	
Contrib'n to Gen. Fund..$4,137,602 73		
Do. Gen'l fund debt 4,234,416 66		
	8,372,019 39	
		$124,340,633 07
Balance due.....................		$19,549,525 42
*Received from General Fund.....		1,386,495 88
Canal debt unpaid and unprovided for.............................		$20,936,027 80

The magnitude of the trade and tonnage of the New York Canals will be shown by the following exhibit, giving the tonnage and revenue in each year from 1820 to 1862, inclusive:

	Tonnage.	Toll, all N. Y. C'ls.	Toll rec'd at Buffalo.
1820..........	....	$5,244	....
1821..........	....	23,388	....
1822..........	....	64,072	....
1823..........	....	190,685	....
1824..........	...	340,643	....

	Tonnage.	Toll, all N. Y. C'ls.	Toll rec'd at Buffalo.
1825..........	....	$566,279	....
1826..........	....	765,104	....
1827..........	....	850,260	....
1828..........	....	833,444	....
1829..........	....	813,187	....
1830..........	....	1,056,922	....
1831..........	....	1,223,809	....
1832..........	....	1,229,483	....
1833	....	1,462,829	....
1834	....	1,341,829	....
1835..........	....	1,545,986	....
1836..........	1,810,807	1,614,836	$106,218
1837..........	1,171,296	1,292,623	....
1838..........	1,183,011	1,590,911	....
1839..........	1,445,713	1,616,882	....
1840..........	1,416,046	1,775,747	....
1841..........	1,521,661	2,034,882	....
1842..........	1,236,931	1,749,196	....
1843..........	1,513,439	2,031,590	....
1844..........	1,810,556	2,446,874	....
1845..........	1,985,011	2,646,181	....
1846..........	2,265,662	2,756,106	....
1847..........	2,869,810	8,635,851	...
1848..........	2,796,230	8,252,212	1,216,701
1849..........	2,894,782	8,264,226	672,614
1850..........	3,076,617	8,275,899	757,491
1851..........	3,582,733	8,329,717	703,498
1852..........	8,863,441	8,118,244	777,102
1853..........	2,247,852	8,204,719	802,657
1854..........	4,165,862	2,773,566	695,897
1855..........	4,022,617	2,805,077	685,810
1856..........	4,116,082	2,948,203	755,905
1857..........	4,344,061	2,045,644	598,470
1858..........	3,665,192	2,110,754	719,683
1859..........	3,751,654	1,723,945	552,432
1860..........	4,650,214	3,009,597	1,137,815
1861..........	4,507,635	8,908,785	2,101,635
1862..........	5,598,785	5,188,943	3,054,058

A comparison of the tonnage and revenue of the New York canals in 1861 and 1862, will show an augmentation in revenue of $1,280,158 from tolls, while the tonnage was increased 1,091,150 tons.

The aggregate tons carried on the New York canals since they first went into operation, will vary but little from eighty-two millions of tons, valued at upwards of four thousand millions of dollars, from which the State has received a gross revenue of upwards of eighty-one millions.

There is, probably, not another system of public works in any country on the face of the globe that has produced in so short a period of time such stupendous results.

There are other commercial interests of great magnitude not mentioned in the above extracts from the Report of the Trade and Commerce of Buffalo, of which we might enumerate different kinds of grain, beef, pork, butter, cheese, whiskey, alcohol, hides and leather, coal, wool, potatoes, fruit, &c.

The trade with Canada, and with foreign ports, passing through the Welland Canal, and down the St. Lawrence River to the Atlantic, is of great and growing importance—destined, no doubt, to increase immensely on the opening of the proposed *Ship Canal,* to connect with the Mississippi River, terminating at Chicago, Illinois, and other internal improvements. A railroad route from the Upper Mississippi, commencing near St. Paul, Minnesota, has been surveyed, and will, no doubt, terminate at Superior City, or Bayfield, favorably situated on the South Shore of Lake Superior.

The incalculable advantages of this latter communication, to Lake Commerce, can only be realized when fully completed—thus draining Northern Iowa, Minnesota, and Dacotah of their rich agricultural products—all of which will flow eastward to the Atlantic, and European markets.

TRADE AND COMMERCE OF CHICAGO.

Extract from the REPORT *of the Committee on Statistics,* 1863.

As an evidence of the increase of agricultural products since 1859, consequent on improved crops and an enlarged area of cultivation, your Committee would direct attention to the provision trade of Chicago for the last four years.

TABLE, SHOWING THE RECEIPTS AT CHICAGO OF THE ARTICLES NAMED FOR THE YEARS 1859-'62.

ARTICLES.	1859.	1860.	1861.	1862.
Flour, barrels......	726,321	713,348	1,479,284	1,666,391
Wheat, bushels....	8,060,766	14,427,083	17,385,002	13,978,116
Corn, "	5,401,870	15,262,394	26,369,989	29,574,328
Oats, "	1,757,696	2,198,889	2,067.018	4,688,722
Rye, "	231,514	318,976	490.989	1,038,825
Barley "	652,696	617,619	457,589	872,053
Hogs............	271,204	392,864	675,902	1,348,890
Cattle........ ...	111,694	177,101	204,579	209,655

Thus the increase in cereals has been 196 per cent.; in hogs, 400 per cent.; and in cattle, 87 per cent.

Results equally marked are shown by the returns of the other lake-ports.

The committee of the Chicago Board of Trade, in a recent Report, say:

"In the early settlement of the West, the Mississippi was the only outlet for the products of the country; but the opening of the New York and Canadian canals, and of not less than five trunk railways between the East and West, has rendered the free navigation of the Mississippi a matter of secondary importance.

"The heated waters of a tropical sea, destructive to most of our articles of export, a malarious climate, shunned by every Northerner for at least one-half of the year, and a detour in the voyage of over 3,000 miles in a direct line to the markets of the world.—these considerations have been sufficiently powerful to divert the great flow of animal and vegetable food from the South to the East. Up to 1860, the West found a local market for an inconsiderable portion of her breadstuffs and provisions in the South; but after supplying this local demand, the amount which was exported from New Orleans was insignificant, hardly exceeding two millions of dollars per annum."

The annual report of the Secretary of the Treasury for the year ending August 31, 1860, shows the amount of breadstuffs and provisions exported to foreign

FLOUR. bbls.	WHEAT. sacks and bbls.	CORN. sacks and bbls.	OATS. sacks and bbls.
965,860	339,348	1,722,637	659,550

These facts show conclusively that, with the navigation of the Mississippi unobstructed, the great mass of Western

countries from New Orleans and New York respectively, as follows:

	From New Orleans.	From New York.
Wheat, bushels	2,189	1,880,908
Wheat Flour, barrels	80,541	1,157,200
Indian Corn, bushels	224,882	1,880,014
Indian Meal, barrels	154	86,073
Pork, barrels	4,259	109,379
Hams and bacon, pounds	890,230	16,161,749

The total receipts of grain of all kinds, at that port, in no single year exceeded 14,500,000 bushels, either for exportation or consumption in the interior, which are about the receipts at Milwaukee, or Toledo. In 1859–'60, the receipts were as follows:

exports would flow through other channels.

PRODUCT OF BREADSTUFFS FOR EXPORTATION.

The amount of cereals, which, in 1862, flowed out of the Upper Mississippi Valley and the region of the Lakes, *en route* for the sea-board, was, according to the Buffalo Trade Report, 136,329,542 bushels, which were respectively forwarded from the following points:

STATEMENT SHOWING THE SHIPMENT OF CEREALS FOR 1862.

Places.	Flour. bbls.	Wheat. bush.	Corn. bush.	Other Grain bush.
W. Terminus B. & O. R. R.*	690.000			550,000
" Pennsylvania Central	890,696			1,622,893
Dunkirk	1,095,365	112,061	149,654	10,173
Suspension Bridge*	875,000			2,750,000
Buffalo	2,846,022	30,435,831	24,288,627	3,849,620
Oswego	235,382	10,982.132	4,528,952	1,467.823
Cape Vincent	48,576	316,403	249,369	49.017
Ogdensburgh	576,394	689,930	1,120,176	18,865
Montreal ▲	1,101,475	8,012.773	2,649,136	519,896
Rochester*	1,000	150,000		6,622
TOTALS	8,359,910	50,699,130	32,985,923	10,914,939
GRAND TOTAL—(Flour reduced to bushels)				136,329,542

* Estimated.

SHIPMENTS OF CEREALS FROM FOUR LAKE PORTS, IN 1862.

Places.	Flour. bbls.	Wheat. bush.	Corn. bush.	Other Grain. bush.
Chicago.....................	1,739,849	13,808,898	29,452,610	4,516,357
Milwaukee	711.405	14,915,680	9,489	250,292
Toledo*.....................	1,261,291	9,314,491	3,781,634	....
Detroit†	998,535	3,278,033	310,618	122,109
TOTALS.................	4,711,080	41,317,102	33,554,351	4,883,758
GRAND TOTAL—(Flour reduced to bushels)				103,315,611

The mining population of Lake Superior absorb not less than 150,000 bushels of cereals, which do not appear in the above tables, and which will account for the discrepancies between the amounts shipped from the initial points, and the amounts forwarded from the secondary points. These tables are illustrative, as showing that, in this great grain-movement, the four lake ports furnish more than fifty per cent. of all the flour, more than eighty per cent. of all the wheat, and more than seventy-five per cent. of the cereals of all kinds; while Chicago and Toledo together furnish more corn than finds its way eastward through all these avenues, and Chicago alone contributes more than forty per cent. of the whole gross product.

STATEMENT,

Showing the Capacity of our Warehouses for Handling and Storing Grain in Chicago.

ELEVATING WAREHOUSES.	Capacity for Storage. bush.	Capacity to receive and ship per day. bush.	Capacity to ship per day. bush.
Sturges, Buckingham & Co., A........ ...	700,000	65,000	225,000
" " B............	700,000	65,000	225,000
Flint & Thompson.....................	160,000	25,000	50,000
" R. I. R. R.............	700,000	55,000	200,000
Charles Wheeler & Co., G. & C. U. R. R....	500,000	50,000	125,000
Munger & Armour ....:	600,000	50,000	100,000
Hiram Wheeler........................	450,000	60,000	150,000
Munn & Scott........................	200,000	30,000	75,000
O. Lunt & Brother.....................	80,000	30,000	40.000
Ford & Norton........................	100,000	40,000	45,000
George Sturges & Co., Fulton Elevator.....	100,000	25,000	50,000
Walker, Washburn & Co..................·	75,000	30,000	60,000
Albert, Sturges & Company...............	700,000	65,000	225,000
Armour, Dole & Co.	850,000	85,000	225,000
Munn & Scott (new house)...............	600,000	55,000	200,000
L. Newberry & Co......................	300,000	40,000	100,000
Flint & Thompson (new house)...........	1,000,000	90,000	230,000
Armour, Dole & Co. "	800,000	90,000	230,000
Total Capacity of Warehouses......	8,615,000	950,000	2,555,000

* Amount received from Chicago deducted.
† Amount received from Chicago and Milwaukee deducted.

PORK AND BEEF PACKING.

The progress which has been made in Pork Packing in Chicago during the past two years, is probably without a parallel in the history of any other city in the United States.

During the past year there have been erected along the River seven large Pork and Beef houses, all of which have been constructed on the most approved plans. Besides these, there have been built a large number of smaller structures, of more or less permanence; all of which, with the temporary occupation of stores, outhouses, &c., give great additional facilities in the extension of this business. As the season is not yet closed, we can only judge of the packing for 1862-'3, by the number cut from the commencement of the season, till January 1, which foots up 539,216, against 229,850, packed during the same period in 1861—an increase, thus far, of 309,366.

During the past two seasons, a large proportion of the Hogs cut have been made into English Middles, for the Liverpool and London markets. In the early part of this season, nearly every packing house in the city was engaged in this branch of the business. The favor with which Chicago brands have been received in the leading markets of England, warrants us in the belief that the trade will be one of permanence.

TABLE

Showing the number of Hogs *Received and Forwarded for five years.*

RECEIVED.

YEAR.	Live.	Dressed.	TOTAL.
1858	416,225	124,261	540,456
1859	188,671	82,583	271,204
1860	285,149	107,515	892,564
1861	549,089	126,868	675,902
1862	1,110,971	237,919	1,348,690

FORWARDED.

YEAR.	Live.	Dressed.	TOTAL.
1858	159,181	32,832	192,018
1859	87,254	22,992	110,244
1860	191,931	85,283	227,164
1861	216,962	72,112	289,084
1862	446,506	44,629	491,135

BEEF CATTLE.

The past year has shown but a small increase in the Cattle Trade of Chicago.

By the tables which follow, it will be seen that the receipts of Cattle at this point amount to 209,655 against 204,579 in 1861, an increase of 5,076 head; and the shipments to 112,745 against 124,146, in the same period of time. Showing a decrease of 11,401.

TABLE

Showing the number of CATTLE *Received and Forwarded for five years.*

Received in 1858	140,584
" 1859	111,604
" 1860	177,101
" 1861	204,579
" 1862	209,655
Forwarded in 1858	42,683
" 1859	87,584
" 1860	97,474
" 1861	124,146
" 1862	112,745

The cereals and agricultural products shipped from Chicago consist of corn, wheat, rye, oats, barley, butter, cheese, potatoes, wool, hides, &c. The products of the forest are lumber, and wood of different kinds. The minerals are coal, &c.; while fisheries furnish large quantities of cured fish of different kinds for exportation; altogether giving employment to a large amount of tonnage navigating the great lakes.

TRIP THROUGH THE LAKES,

Giving a Description of Cities, Towns, &c.

ing West. The harbor of Buffalo is the most capacious, and really the easiest and safest of access on our inland waters. Improvements are annually made by dredging, by the construction of new piers, wharves, warehouses, and elevators, which extend its facilities, and render the discharge and trans-shipment of cargoes more rapid and convenient; and in this latter respect it is without an equal.

Buffalo, "QUEEN CITY of the LAKES," possessing commanding advantages, being 22 miles above Niagara Falls, is distant from Albany 298 miles by railroad, and about 350 miles by the line of the Erie Canal; in N. lat. 42° 53′, W. long. 78° 55′ from Greenwich. It is favorably situated for commerce at the head of Niagara River, the outlet of Lake Erie, and at the foot of the great chain of Upper Lakes, and is the point where the vast trade of these inland seas is concentrated. The harbor, formed of Buffalo Creek, lies nearly east and west across the southern part of the city, and is separated from the waters of Lake Erie by a peninsula between the creek and lake. This harbor is a very secure one, and is of such capacity, that although steamboats, ships, and other lake craft, and canal-boats, to the number, in all, of from three to four hundred, have sometimes been assembled there for the transaction of the business of the lakes, yet not one-half part of the water accommodations has ever yet been occupied by the vast business of the great and grow-

Buffalo was first settled by the whites in 1801. In 1832 it was chartered as a city, being now governed by a mayor, recorder, and board of twenty-six aldermen. Its population in 1830, according to the United States Census, was 8.668; in 1840, 18,213; and in 1850, 42,261. Since the latter period the limits of the city have been enlarged by taking in the town of Black Rock; it is now divided into thirteen wards, and, according to the Census of 1860, contained 81,130 inhabitants, being now the third city in point of size in the State of New York. The public buildings are numerous, and many of them fine specimens of architecture; while the private buildings, particularly those for business purposes, are of the most durable construction and modern style. The manufacturing establishments, including several extensive ship-yards for the building and repairing of lake craft, are also numerous, and conducted on a large scale, producing manufactured articles for the American and Canadian markets.

The principal public buildings are an

U. States Custom-House and Post-Office; City Hall; Court-House and Jail; 2 Theatres, and 50 Churches of different denominations. Here are also 8 banking houses, 4 Savings Banks, and several Fire and Marine Insurance Companies.

The Lines of Steamers and Railroads diverging from Buffalo tend to make it one of the greatest thoroughfares in the Union. Steamers and propellers run to Cleveland, Sandusky, Toledo, Detroit, Mackinac, Saut Ste. Marie, Green Bay, Milwaukee, Chicago, &c.

RAILROADS RUNNING FROM BUFFALO.

1. *New York Central*, to Albany and Troy, 298 miles.
2. *Buffalo, Niagara Falls, and Lewiston*, 28 miles.
3. *Buffalo, New York, and Erie*, to the city of New York, 434 miles.
4. *Lake Shore Railroad*, to Cleveland, Ohio, 183 miles.
5. *Buffalo and Lake Huron Railway*, 161 miles.
6. *Great Western* (Canada) *Railway* (Sus. Bridge to Detroit, Mich.). 230 miles.

There are also four lines of City Railroads running to different points within the limits of Buffalo.

The principal Hotels are the *American*, and *St. James Hotel*, and *Mansion House*, on Main street; *Bonney's Hotel*, on Washington street, and the *United States*, facing the Terrace.

There are now running from Buffalo to different ports on Lake Erie and the Upper Lakes seven different Lines of Propellers, annually transporting an immense amount of merchandise and produce, which finds its exit eastward by means of the Erie Canal, and the several Railroad Lines.

"The climate of Buffalo is, without doubt, of a more even temperature than any other city in the same parallel of latitude from the Mississippi to the Atlantic coast. Observations have shown that the thermometer never ranges as low in winter, nor as high in summer, as at points in Massachusetts, the eastern and central portions of this State, the northern and southern shores of Lake Erie in Michigan, Northern Illinois, and Wisconsin. The winters are not as keen, nor the summers, cooled by the breezes from the lake, as sultry; and in a sanitary point of view, it is probably one of the healthiest cities in the world.

"London, usually considered the healthiest of cities, has a ratio of one death in forty inhabitants. The ratio of Buffalo is one in fifty-six. The favorable situation of the city for drainage, and for a supply of pure water; its broad, well-paved streets, lined with shrubbery and shade-trees; its comparatively mild winters; its cool summers; its pleasant drives and picturesque suburbs, and its proximity to the 'Falls,' combine to render it one of the most desirable residences on the continent."

BUFFALO TO DETROIT—NORTH SHORE ROUTE.

On leaving the wharf at Buffalo, the Steamers usually run direct for Long Point on the Canada, or North Shore of Lake Erie, proceeding for most of the distance in British waters, to the mouth of Detroit River.

LONG POINT, 65 miles from Buffalo, is a long strip of land, nearly 20 miles long, and from one to three miles in width, covered for the most part with a stunted growth of forest trees. It was formerly a peninsula, running out from the land in an easterly direction, nearly half way across the lake; but the waters having

48 TRIP THROUGH THE LAKES.

made a wide breach across its western extremity, has converted it into an island. There is an important light-house on the east end to guide the mariner on his passage through Lake Erie, here about 40 miles wide, and where is found the greatest depth of water. To this Point both shores of the lake can be seen in a clear morning from the deck of the steamer, affording a most grand sight when the sun rises on a cloudless day. Then may usually be seen a fleet of vessels wending their way toward Buffalo or the mouth of the Welland Canal, through which channel annually passes a great number of steam propellers and sail vessels on their way to Lake Ontario and the St. Lawrence River.

PORT COLBORNE, C. W., situated about 20 miles west of Buffalo, lies at the mouth of the Welland Canal, while

PORT MAITLAND, some 20 miles farther, is situated at the mouth of Grand River, where is a navigable feeder communicating with the canal, thus affording two entrances to the above canal.

PORT DOVER, about 70 miles west of Buffalo, and 40 miles distant from Hamilton by proposed railroad route, is situated on the north shore of Lake Erie, at the mouth of the river Lynn. Here is a good harbor, and the village is a place of growing importance, containing about 1,000 inhabitants.

PORT RYERSE and PORT ROWAN are small villages on the Canada shore, situated on the bay formed by Long Point. Inland there is to be found a rich and fine farming district, consisting of some of the best lands in Canada West.

The *Sand Hills*, immediately west of Long Point, are seen for some distance as the steamer pursues her onward course toward *Point aux Pins*, passing through the widest part of the lake, where both shores are lost sight of for a number of miles. The water usually presents a clear green color in the middle, but near the shore is more or less tinged with muddy water, proceeding from the streams emptying into the lake.

PORT BURWELL, C. W., about 35 miles west of Long Point, is handsomely situated at the mouth of Otter Creek. Here is a light-house and good harbor. A large amount of lumber and other products are annually exported from this place to Eastern markets.

PORT STANLEY, about 25 miles farther west, is handsomely situated at the mouth of Kettle Creek, being in part surrounded by high and picturesque hills in the immediate vicinity. The harbor is well protected, and much frequented by British and American vessels running on Lake Erie. It is nine miles south of St. Thomas, and twenty-four from London, the chief town of the county of Middlesex, for which place it may be considered the out-port. A plank-road runs between the two places; also, the *London and Port Stanley Railroad*, connecting with the Great Western Railway of Canada. Steamers run from Port Stanley to Buffalo, Cleveland, and other ports on Lake Erie.

POINT AUX PINS, or ROND' EAU (usually called by the American navigators *Round O*), about 100 miles west of Long Point, is a cape which projects from the Canada shore, enclosing a natural basin of about 6,000 acres in extent, with a depth of from ten to twelve feet, thus forming an excellent and secure harbor, the entrance to which has been improved by the Canadian government by running out piers, etc. It is proposed to construct a ship canal from this port to the St. Clair River, a distance of about 35 miles, thus avoiding the *St. Clair Flats*. Another Canadian project is to construct a canal from Goderich to Hamilton, C. W., about 120 miles in length.

POINT PELEE, lying about 40 miles east of the mouth of Detroit River, projects a number of miles into Lake Erie, and forms, in connection with the island of Point Pelee and other islands in the vicinity, the most picturesque portion of lake scenery to be met with on this inland sea.

POINT PELÉE ISLAND, belonging to Canada, is about seven miles long, and two or three miles in width. It is inhabited by a few settlers. The island is said to abound with red cedar, and possesses a fine lime-stone quarry. A lighthouse is situated on the east side.

The steamers bound for Detroit River usually pass to the north side of Point Pelee Island, and run across *Pigeon Bay* toward *Bar Point*, situated at the mouth of Detroit River. Several small islands are passed on the south, called *East Sister, Middle Sister*, and *West Sister;* also, in the distance, may be seen the BASS ISLANDS, known as the "North Bass," "Middle Bass," and "South Bass." On the west side of the latter lies the secure harbor of PUT-IN-BAY, celebrated as the rendezvous of Com. Perry's fleet, before and after the glorious naval victory which he achieved over the British fleet, September 10th, 1813.

DETROIT RIVER, forming one of the links between the Upper and Lower Lakes, is next approached, near the mouth of which may be seen a light on the Michigan shore called *Gibraltar Light*, and another light on an island attached to Canada, the steamers usually entering the river through the east or *British Channel* of the river, although vessels often pass through the west or *American Channel*.

AMHERSTBURGH, C. W., 18 miles below Detroit, is an old and important town. The situation is good; the banks of the river, both above and below the village, but particularly the latter, where the river emerges into Lake Erie, are very beautiful; several handsome residences

may here be seen, surrounded by highly cultivated grounds. About a mile below the town is a chalybeate spring, which is said to resemble the waters of Cheltenham, in England. British and American vessels frequently land at Amherstburg, on their trips to and from the Upper Lakes.

FORT MALDEN, capable of accommodating a regiment of troops, is situated about half a mile above Amherstburg, on the east bank of the river, the channel of which it here commands.

At BROWNSTOWN, situated on the opposite side of the river, in Michigan, is the *battle-ground* where the Americans, under disadvantageous circumstances, and with a slight loss, routed the British forces, which lay in ambush, as the former were on their way to relieve the fort at Frenchtown, which event occurred August 5, 1812.

SANDWICH, C. W., is beautifully situated on the river, two miles below Detroit, and nine miles below Lake St. Clair. It stands on a gently sloping bank a short distance from the river, which is here about a mile wide. This is one of the oldest settlements in Canada West. The town contains 3,133 inhabitants.

WINDSOR, C. W., situated in the township of Sandwich, is a village directly opposite Detroit, with which it is connected by three steam ferries. It was laid out in 1834, and is now a place of considerable business, having a population of about 2,500 inhabitants. Here terminates the *Great Western Railway* of Canada, which extends from Niagara Falls or Suspension Bridge, *via* Hamilton and London, to opposite Detroit—thus forming an important link in the great line of railroads, now finished, running from the seaboard at different points to the Mississippi and Missouri Rivers.

4

RAILROAD ROUTE from Niagara Falls to Hamilton and Detroit, *via* Great Western Railway of Canada.

This great International Line, extending from Niagara River to Detroit River, opposite the city of Detroit, a distance of 229 miles, passes through a fine and interesting section of country, equal in many respects to Western New York. It connects with the New York Central and Buffalo and Niagara Falls Railroad, forming a great through route of travel.

Starting from the *Suspension Bridge* at Clifton, two miles below the Falls of Niagara, the passenger train soon reaches the verge of the mountain ridge overlooking the plain below, while in the distance may be seen the broad waters of Lake Ontario, usually studded with sail vessels and propellers on their way to or from the mouth of the Welland Canal.

> "Traced like a map, the landscape lies
> In cultured beauty stretching wide."

THOROLD, nine miles, is situated on the line of the Welland Canal, where is abundant water-power propelling five or six flouring mills. A railroad extends to Port Dalhousie, some five or six miles distant, connecting with a steamer running to Toronto. This road will be extended to Port Colbourne, on Lake Erie, about twenty miles distant.

ST. CATHERINES, 12 miles from the Suspension Bridge, is a flourishing town, also situated on the line of the Welland Canal, which connects Erie and Ontario. This has become of late a fashionable place of resort during the summer months, caused by the mineral waters of the "*Artesian Wells*" obtaining great celebrity, owing to their curative properties. Here are several well-kept hotels for the accommodation of visitors. St. Catherines is justly termed "*the Saratoga of Canada,*" being annually visited by thousands of invalids and pleasure-seekers.

BEAMSVILLE, 22 miles from the Suspension Bridge, is a thriving village about one mile from the station.

GRIMSBY, 5 miles farther, is situated on Forty-mile Creek, the scene of some hard fighting during the war of 1812. It is a small village of 350 inhabitants; there are two churches, a hotel, and several stores; also, a grist and saw mills propelled by water-power.

Hamilton, 43 miles from Suspension Bridge, is the principal station on the line of the Great Western Railway, where are located the principal offices and workshops connected with the company. Here is a commodious dépôt and steamboat landing. Carriages and omnibuses are always in readiness to convey passengers to the hotels in the city, and steamboat landings.

The *Toronto Branch* of the Great Western Railway commences at Hamilton, and extends a distance of thirty-eight miles to the city of Toronto, running near the shore of Lake Ontario.

On leaving Hamilton for Windsor or Detroit, the road passes near the mansion of the late Sir Allan M'Nab, and over the Des Jardines Canal, entering the head of Burlington Bay. Here is also a Suspension Bridge in sight, thrown over the stream as it cuts its way through the high bank which encircles the bay or lake. This point presents a beautiful view, both on leaving or arriving at the head-waters of Lake Ontario.

DUNDAS, five miles from Hamilton, is situated on rising ground on the side of the mountain, and is a thriving manufacturing place, having the advantage of a stream which flows, or rather rushes, with great impetuosity through its centre, working on its way numerous mills. The *Des Jardines Canal* runs from hence to Burlington Bay, enabling the manufacturers to ship their goods at their own doors.

HARRISBURGH, 19 miles from Hamilton, is the station of the *Galt Branch* of the Great Western Railway.

PARIS, with the Upper and Lower Town, contains about 3,500 inhabitants; so called from its contiguity to beds of gypsum or plaster of Paris. It possesses a considerable amount of water-power, which works numerous mills. There are two foundries, a tannery, machine-shop, distillery, saw-mill, etc. The *Buffalo and Lake Huron Railway* intersects the Great Western at this point, running to Goderich, on Lake Huron.

WOODSTOCK, 43 miles from Hamilton, and 138 miles from Windsor, is a county town, well situated on rolling ground, and contains about 4,500 inhabitants. It may be called a town of magnificent distances; East and West Woodstock forming a street upward of a mile in length. The vacant spaces, however, are fast being filled up with stately edifices, and it will thus in a short time become one of the handsomest thoroughfares in Canada. In this locality, noted for its handsome country seats—and indeed all the way from Hamilton—the land, as seen from the road (the railroad for the most part passes through a new country), is rolling and well cleared of trees and stumps, presenting more the appearance of "merrie England" than any other section of the Province.

INGERSOLL, nine miles farther, formerly an Indian village, now contains about 2,000 inhabitants. A small arm of the Thames runs through it, and furnishes some water-power, by which several mills are worked. Since the opening of the railway it has risen in a surprising manner; and the town, which before then had a very dingy appearance, the houses being of wood and wanting paint, is now gay with white brick, and the streets resound with the hum of an enterprising population.

LONDON, 119 miles from Suspension Bridge, and 110 miles from Windsor, if not, like her English namesake,

————The great resort
Of all the earth—checkered with all
Complexions of mankind—

is nevertheless a very stirring business place, and presents another instance of the energy and enterprise of the Canadians. Ten years ago, this then very small village of wooden houses was entirely burned down, and now on its ashes is raised a most flourishing city, containing four banks, several wholesale houses, fifteen churches, many of them handsome structures, and the English Church having a fine peal of bells; life and fire insurances offices, breweries and distilleries. It has three newspapers and several good hotels. Population nearly 18,000. It is well watered by the river Thames, which, however, is only navigable up to Chatham, sixty miles distant.

The *London and Port Stanley R. R.* here joins the Great Western Railway; length 24 miles, running south to Lake Erie.

CHATHAM, 46 miles from Windsor, situated on the river Thames, possesses the great advantage of a navigation, and is therefore a place of considerable business. It contains eight churches; and being the county town of Kent, it has a court-house, a very handsome building, several grist and saw mills, woollen factory, two foundries, machine-shop, etc. Steamers ply between Chatham, Detroit, and Amherstburg. Population about 5,000.

WINDSOR, 229 miles from Suspension Bridge, opposite Detroit, prettily situated on the banks of the river, is a place of considerable business, and is rapidly increasing in wealth and population, owing to the advantage it has of being the western terminus of the Great Western Railway.

Three steam-ferries ply between Windsor and Detroit, making close connections for the benefit of railroad passengers.

BUFFALO TO GODERICH, C. W., *via* BUFFALO AND LAKE HURON RAILWAY.

This important line of travel extends from Buffalo, N. Y., crossing Niagara River by means of a steam ferry at Black Rock to Fort Erie, on the Canada side. It is proposed to construct a permanent railroad bridge of about one mile in length, a short distance above the present ferry. From Fort Erie the line of the railway extends westward within a short distance of Lake Erie for forty miles, to Dunville, situated at the mouth of Grand River, crossing the Welland Canal.

From Dunville the road runs along the valley of the river on the north side to Brantford, 38 miles farther, and from thence extends westward to Paris, where it connects with the Great Western Railway of Canada. The line thence runs to Stratford, C. W., where it connects with the Grand Trunk Railway, a total distance from Buffalo of 116 miles. From this point the road is now completed and in running order to Goderich, situated on Lake Huron, a total distance of 163 miles.

DUNVILLE is advantageously situated on the Grand River, at a point where it is intersected by the feeder of the Welland Canal. It is a place of considerable business, and contains several grist, saw, and plaster mills, and a tannery. Population about 1,500.

The *Welland Canal* is one of the many works of the same kind of which Canadians may be proud. This Canal affords a passage for propellers, sloops, and schooners of 125 tons burden, around the Falls of Niagara, and connects Lake Erie with Lake Ontario. It is 42 miles long, including feeder, 56 feet wide, and from 8¼ to 16 feet deep. The whole descent from one lake to the other is 334 feet, which is accomplished by 37 locks.

BRANTFORD, 78 miles from Buffalo, and 82 miles from Goderich, is beautifully situated on Grand River, and named after Brant, the renowned chief of the Six Nations Indians, who, with his tribe, steadily supported the British Crown during the American War. "In '*Gertrude of Wyoming*' he is alluded to in disparaging terms:

'The mammoth comes—the fiend, the monster Brant.'

But some years afterward Campbell was obliged to apologize to Brant's son, who happened to visit London; as it appeared, on satisfactory evidence, his father was not even present at the horrible desolation of Wyoming. This much is due to the memory of Brant, who was a brave warrior and a steadfast ally of the British, and always exerted himself to mitigate the horrors of war."

Brantford, until the opening of the Great Western Railway, was a great wheat market, the streets being crowded with hundreds of wagons daily; but that road created other markets, and to this extent the town has suffered. It has, however, other sources of prosperity. There is no place in the Province which commands such extensive water-power, and which is made available for the working of numerous mills. The iron foundries, machine shops, and potteries are on a large scale, and have caused the place to be regarded as the Birmingham of Canada. It has a goodly number of churches of various denominations, and one of the largest and handsomest hotels in the Province—"The Kirby House." Population about 6,000.

STRATFORD is a new and thriving town, favorably situated on the line of the *Grand Trunk Railway* of Canada. This section of Canada enjoys a good climate and fertile soil, producing cereal grains in great abundance.

Goderich, C. W., 163 miles distant from Buffalo, by railroad route, is advantageously situated on the east shore of Lake Huron, enjoying a healthy and delightful climate. Here is a good and secure harbor, being easily accessible to the largest steamers and sail vessels navigating the Upper Lakes. Here is erected an extensive railroad depot, warehouses, grain elevator, and wharves, owned by the Buffalo and Lake Huron Railway company. The town is beautifully situated on elevated ground, rising about 150 feet above the waters of Lake Huron, here celebrated for their purity. The population amounts to about 5,000, and is rapidly increasing in numbers and wealth. Steamers run daily from this port to Sarnia, Detroit, Saginaw, and other ports on Lake Huron. A line of propellers, carrying passengers and freight, also run from Goderich to Mackinac, Milwaukee, Chicago, etc.—connecting with the Buffalo and Lake Huron Railway—thus forming a through line of travel from Buffalo to the above ports.

There are several other ports of importance lying north of Goderich, on the Canada side of the lake, from whence steamers run to and fro almost daily, during the season of navigation.

TABLE OF DISTANCES.

FROM BUFFALO TO TOLEDO—SOUTH SHORE ROUTE.

Ports, etc.	Miles.	Ports, etc.	Miles.
Buffalo, N. Y	0	Toledo, Ohio	0
Silver Creek, N. Y	34	Maumee Bay	7
Dunkirk, "	42	Turtle Island	10
Portland, "	52	West Sister Island	22
Erie, Pa	90	South Bass Island	40
Conneaut, Ohio	117	Sandusky, Ohio	40
Ashtabula, "	131	Kelley's Island	45
Painesville, "	156	Cleveland, Ohio	100
Cleveland, "	185	Painesville, "	129
Kelley's Island	240	Ashtabula, "	154
Sandusky, Ohio	245	Conneaut, "	168
South Bass Island	245	Erie, Pa	195
West Sister Island	263	Portland, N. Y	233
Turtle Island	275	Dunkirk, "	243
Maumee Bay	278	Silver Creek, N. Y	251
Toledo, Ohio	285	Buffalo, N. Y	285

NOTE.—The direct through route as run by the steamers from Buffalo to Toledo is about 250 miles; the circuit of Lake Erie being about 560 miles.

BUFFALO TO CLEVELAND, TOLEDO, ETC.—SOUTH SHORE ROUTE.

Steamers •and propellers of a large class leave Buffalo daily, during the season of navigation, for the different ports on the American or South Shore of Lake Erie, connecting with railroad cars at Cleveland, Sandusky, Toledo, and Detroit.

On leaving Buffalo harbor, which is formed by the mouth of Buffalo Creek, where is erected a breakwater by the United States government, a fine view is afforded of the city of Buffalo, the Canada shore, and Lake Erie stretching off in the distance, with here and there a steamer or sail vessel in sight. As the steamer proceeds westward through the middle of the lake, the landscape fades in the distance, until nothing is visible but a broad expanse of green waters.

STURGEON POINT, 20 miles from Buffalo, is passed on the south shore, when the lake immediately widens by the land receding on both shores. During the prevalence of storms, when the full blast of the wind sweeps through this lake, its force is now felt in its full power, driving the angry waves forward with the velocity of the race-horse, often causing the waters to rise at the lower end of the lake to a great height, so as to overflow its banks, and forcing its surplus waters into the Niagara River, which causes the only perceptible rise and increase of the rush of waters at the Falls. •

DUNKIRK, N. Y., 42 miles from Buffalo, is advantageously situated on the shore of Lake Erie where terminates the *New York and Erie Railroad*, 460 miles in length. Here is a good and secure harbor, affording about twelve feet of water over the bar. A light-house, a beacon-light, and breakwater, the latter in a dilapidated state, have here been erected by the United States government. As an anchorage and port of refuge this harbor is extremely valuable, and is much resorted to for that purpose by steamers and sail vessels during the prevalence of storms.

The village was incorporated in 1837, and now contains about 4,000 inhabitants, 500 dwelling-houses, five churches, a bank, three hotels, and 20 stores of different kinds, besides several extensive store-houses and manufacturing establishments.

The *Buffalo and State Line Railroad*, extending to Erie, Pa., runs through Dunkirk, forming in part the Lake Shore line of railroad, which, in connection with the railroad leading direct to the city of New York, affords great advantages to this locality, which is no doubt destined to increase with the growing trade of the lake country.

FREDONIA, three miles from Dunkirk, with which it is connected by a plank-road, is handsomely situated, being elevated about 100 feet above Lake Erie. It contains about 2,300 inhabitants, 300 dwelling-houses, five churches, one bank, an incorporated academy, four taverns, twenty stores, besides some mills and manufacturing establishments situated on Canadoway Creek, which here affords good water-power. In the village, near the bed of the creek, is an inflammable spring, from which escapes a sufficient quantity of gas to light the village. A gasometer is constructed which forces the gas through tubes to different parts of the village, the consumer paying $4 per year for each burner used. It is also used for lighting the streets of the village. The flame is large, but not so strong or brilliant as that obtained from gas in our cities; it is, however, in high favor with the inhabitants.

BARCELONA, N. Y., 58 miles from

Buffalo, is the westernmost village in the State. It is a port of entry, and is much resorted to by steamers and large vessels navigating the lake, affording a tolerably good harbor, where is situated a light-house which is lighted by inflammable gas; it escapes from the bed of a creek about half a mile distant, and is carried in pipes to the light-house.

Erie, "THE LAKE CITY OF PENNSYLVANIA," distant 90 miles from Buffalo and 95 miles from Cleveland, is beautifully situated on a bluff, affording a prospect of Presque Isle Bay and the Lake beyond. It has one of the largest and best harbors on Lake Erie, from whence sailed Perry's fleet during the war of 1812. The most of the vessels were here built, being finished in seventy days from the time the trees were felled; and here the gallant victor returned with his prizes after the battle of Lake Erie, which took place September 10th, 1813. The remains of his flag-ship, the *Lawrence*, lie in the harbor, from which visitors are allowed to cut pieces as relics. On the high bank, a little distance from the town, are the ruins of the old French fort, Presque Isle. The city contains a court-house, nine churches, a bank, three hotels, a ship-yard, several extensive manufacturing establishments, and about 10,000 inhabitants. In addition to the *Lake Shore Railroad*, the *Philadelphia and Erie Railroad** terminates at this place, affording a direct communication with New York, Philadelphia, and Baltimore.

Presque Isle Bay is a lovely sheet of water, protected by an island projecting into Lake Erie. There is a light-house on the west side of the entrance to the bay, in lat, 42° 8' N.; it shows a fixed light, elevated 93 feet above the surface of the lake, and visible for a distance of 15 miles. The beacon shows a fixed light, elevated 28 feet, and is visible for nine miles.

CONNEAUT, Ohio, 117 miles from Buffalo and 68 from Cleveland, situated in the northeast corner of the State, stands on a creek of the same name, near its entrance into Lake Erie. It exports large quantities of lumber, grain, pork, beef, butter, cheese, etc., being surrounded by a rich agricultural section of country. The village contains about 2,000 inhabitants. The harbor of Conneaut lies two miles from the village, where is a light-house, a pier, and several warehouses.

ASHTABULA, Ohio, 14 miles farther west, stands on a stream of the same name, near its entrance into the lake. This is a thriving place, inhabited by an intelligent population estimated at 3,500. The harbor of Ashtabula is two and a half miles from the village, at the mouth of the river, where is a light-house.

FAIRPORT stands on the east side of Grand River, 155 miles from Buffalo. It has a good harbor for lake vessels, and is a port of considerable trade. This harbor is so well defended from winds, and easy of access, that vessels run in when they cannot easily make other ports. Here is a light-house and a beacon to guide the mariner.

PAINESVILLE, Ohio, three miles from Fairport and 30 miles from Cleveland, is a beautiful and flourishing town, being surrounded by a fine section of country. It is the county seat for Lake County, and contains a court-house, five churches, a bank, 20 stores, a number of beautiful residences, and about 3,000 inhabitants.

* This great line traverses the Northern and Northwest counties of Pennsylvania to the city of Erie on Lake Erie. It has been leased by the Pennsylvania Railroad Company, and under their auspices is being rapidly opened throughout its entire length. It is now in use for passenger and freight business from Harrisburg to Driftwood (177 miles), on the Eastern Division, and from Sheffield to Erie, on the Western Division (75 miles).

is built rises abruptly from the lake level, where stands a light-house, near the entrance into the harbor, from which an extensive and magnificent view is obtained, overlooking the city, the meandering of the Cuyahoga, the line of railroads, the shipping in the harbor, and the vessels passing on the Lake.

The city is regularly and beautifully laid out, ornamented with numerous shade-trees, from which it takes the name of "Forest City." Near its centre is a large public square, in which stands a beautiful marble statue of Commodore OLIVER H. PERRY, which was inaugurated Sept. 10, 1860. in the presence of more than 100,000 people. It commemorates the glorious achievement of the capture of the British fleet on Lake Erie, September 10th, 1813. Cleveland is the mart of one of the greatest grain-growing States in the Union, and has a ready communication by railroad with New York, Boston, and Philadelphia on the east, while continuous lines of railroads run south, and west to the confines of settlement in Kansas and Nebraska. It is distant 185 miles from Buffalo, 135 miles from Columbus, 107 miles from Toledo, and 144 miles from Pittsburgh by railroad route; 120 miles from Detroit by steamboat route.

It contains a County Court-House and Jail, City Hall, U. S. Custom-House and Post Office building; 1 Theatre; a Library Association with a public reading-room; 2 Medical Colleges, 2 Orphan Asylums, 35 Churches of different denominations; 4 Banks, a Savings Bank, and 2 Insurance Companies; also, numerous large manufacturing companies, embracing iron and copper works, ship-building, &c.; Gas-works, Water-works, and two City Railroad Companies. The stores and warehouses are numerous, and many of them well built. It now boasts of 50,000 inhabitants, and is rapidly increasing in numbers and wealth. The Lake Superior trade is a source of great advantage and

Perry Monument, Erected Sept. 10, 1860.

Cleveland, "THE FOREST CITY," Cuyahoga County, Ohio, is situated on a plain, elevated 80 feet above the waters of Lake Erie, at the mouth of the Cuyahoga river, which forms a secure harbor for vessels of a large class; being in N. lat. 41° 30', W. long. 81° 42'. The bluff on which it

profit, while the other lake traffic, together with the facilities afforded by railroads and canals, makes Cleveland one of the favored cities bordering on the Inland Seas of America.

The principal Hotels are the *American Hotel, Angier House, Forest City House, Johnson House,* and *Weddell House;* all being large and well-kept public houses.

RAILROADS DIVERGING FROM CLEVELAND.

1. *Cleveland and Erie,* 95 miles in length.
2. *Cleveland, Columbus, and Cincinnati,* 135 miles.
3. *Cleveland and Toledo,* Northern Division, 107 miles.
4. *Cleveland and Mahoning,* 67 miles finished.
5. *Cleveland and Pittsburgh,* connecting with Wheeling, Va., 200 miles.
6. *Cleveland, Zanesville, and Cincinnati,* 87 miles; diverging from Cleveland and Pittsburgh R. R. at Hudson. These roads all run into one general Depot, situated near the lake, affording great facilities for the trans-shipment of freight and produce of different kinds.

STEAMERS and PROPELLERS of a large class leave daily, during the season of navigation, for Buffalo, Toledo, Detroit, Mackinac, Green Bay, Milwaukee, Chicago, the Saut Ste. Marie, and the different ports on Lake Superior, altogether transporting an immense amount of merchandise, grain, lumber, iron, and copper ore. The registered Tonnage of this port, in 1861, was 82,518 tons.

The *Northern Transportation Company of Ohio* has its principal office in Cleveland. The Company owns 15 propellers of about 350 tons burden, running from Ogdensburgh and Oswego to Cleveland, Toledo, Detroit, Milwaukee, and Chicago. This line affords a cheap and speedy route for travellers and emigrants, as well as for the transportation of merchandise and produce.

The *Cleveland Iron Mining Company,* with a capital stock of $500,000, has its principal office in this city. The mine is situated near Marquette, Lake Superior, being distant about 14 miles from the steamboat landing. A railroad extends to the Iron Mountain, affording facilities for the transportation of 2,000 tons of iron ore per day. This ore yields on an average 66⅓ per cent. of iron. The greater proportion of this ore finds a ready market in Cleveland, from whence the most of it is transported to the Mahoning Valley, where it meets the coal of that region and is smelted and manufactured into merchantable iron.

Steamboat Route from Cleveland to Detroit.

Ports, etc.	Miles.	Ports, etc.	Miles.
CLEVELAND, Ohio	0	DETROIT, Mich	0
Point Pelée Is. and Light	60	Windsor, C. W	1
Bar Point, C. W	97	Fighting Island	8
Bois Blanc Is. Light,		Fish Island	9
Detroit River.	100	Wyandotte, Mich	11
Malden, C. W	101	Mama Juba Is. and Light	12
Gibraltar, Mich		Grosse Isle	18
Grosse Isle, "	102	Gibraltar, Mich	
Mama Juba Is. and Light	108	Malden, C. W	19
Wyandotte, Mich	109	Bois Blanc Is. Light,	
Fish Island Light	111	Lake Erie,	20
Fighting Island	112	Bar Point, C. W	23
Windsor, C. W	119	Point Pelée Island	60
Detroit	120	CLEVELAND	120
		FARE, $3 00. USUAL TIME, 7 hours.	

BLACK RIVER, 28 miles from Cleveland, is a small village with a good harbor, where is a ship-yard and other manufacturing establishments.

VERMILION, 10 miles farther on the line of the Cleveland and Toledo Railroad, is a place of considerable trade, situated at the mouth of the river of the same name.

HURON, Ohio, 50 miles from Cleveland and 10 miles from Sandusky, is situated at the mouth of Huron River, which affords a good harbor. It contains several churches, 15 or 20 stores, several warehouses, and about 2,000 inhabitants.

The islands lying near the head of Lake Erie, off Sandusky, are KELLEY'S ISLAND, NORTH BASS, MIDDLE BASS, and SOUTH BASS islands, besides several smaller islands, forming altogether a handsome group. *Kelley's Island*, the largest and most important, is famous for its grape culture, and has become a place of summer resort by the citizens of Ohio and other States. On the north side of South Bass Island, lies the secure harbor of PUT-IN-BAY, made celebrated by being the rendezvous of Com. Perry's flotilla before and after the decisive battle of Lake Erie, which resulted in the capture of the entire British fleet.

NAVAL BATTLE ON LAKE ERIE.

September 10th, 1813, the hostile fleets of England and the United States on Lake Erie met near the head of the Lake, and a sanguinary battle ensued. The fleet bearing the "red cross" of England consisted of six vessels, carrying 64 guns, under command of the veteran Com. Barclay; and the fleet bearing the "broad stripes and bright stars" of the United States, consisted of nine vessels carrying 54 guns, under command of the young and inexperienced, but brave, Com. Oliver H. Perry. The result of this important conflict was made known to the world in the following laconic dispatch, written at 4 P. M. of that day:

"*Dear General:* We have met the enemy, and they are ours: Two ships two brigs, one schooner, and one sloop. With esteem, etc.,   O. H. PERRY
"Gen. William H. Harrison."

Sandusky, "THE BAY CITY" capital of Erie Co., Ohio, is a port of entry and a place of considerable trade. It is advantageously situated on Sandusky Bay, three miles from Lake Erie, in N. lat. 41° 27', W. long. 82° 45'. The bay is about 20 miles long, and five or six miles in width, forming a capacious and excellent harbor, into which steamers and vessels of all sizes can enter with safety. The average depth of water is from ten to twelve feet. The city is built on a bed of limestone, producing a good building material. It contains about 10,000 inhabitants, a court-house and jail, eight churches, two banks, several well-kept hotels, and a number of large stores and manufacturing establishments of different kinds. This is the terminus of the *Sandusky, Dayton,* and *Cincinnati Railroad,* 153 miles to Dayton, and the *Sandusky, Mansfield,* and *Newark Railroad,* 116 miles in length. The *Cleveland* and *Toledo Railroad,* northern division, also terminates at Sandusky.

Toledo, one of the most favored Cities of the Lakes, is situated on the Maumee river, four miles from its mouth, and ten miles from the Turtle Island Light, at the outlet of the Maumee Bay into Lake Erie. The harbor is good, and the navigable channel from Toledo of sufficient depth for all steamers or sail vessels navigating the lakes. Toledo is the eastern terminus of the *Wabash and Erie Canal,* running through the Maumee and Wabash valleys, and communicating with the Ohio River at Evansville, a distance of 474 miles; also of the *Miami and Erie Canal,* which branches from the above canal 68 miles west of Toledo, and runs southwardly through the Miami

Valley in Western Ohio, and communicates with the Ohio River at Cincinnati, forming together the longest line of canal navigation in the United States.

The railroads diverging from Toledo are the *Michigan Southern and Northern Indiana Railroad*, running through the southern counties of Michigan and the northern counties of Indiana, and making its western terminus at Chicago, Illinois, at a distance of 243 miles; the *Air Line Railroad*, running due west from Toledo, through Northwestern Ohio and the northern counties of Indiana to Goshen, a distance of 110 miles, where it connects with the *Northern Indiana Railroad*, running to Chicago; and the *Detroit, Monroe, and Toledo Railroad*. It is also the eastern terminus of the *Toledo, Wabash, and Western Railroad*, running in a southwesterly direction through the Maumee and Wabash valleys, crossing the eastern line of the State of Illinois, about 125 miles south of Chicago, and continuing in a southwesterly course through Danville, Springfield, Jacksonville, Naples, etc., in Central Illinois, to the Mississippi River, and connecting with the Hannibal and St. Joseph Road, which stretches nearly due west through the State of Missouri to St. Joseph, on the Missouri River. The *Dayton and Michigan Railroad*, which connects Toledo with Cincinnati, is much the shortest railroad line connecting Lake Erie with the Ohio River. Besides the above important roads, the *Cleveland and Toledo Railroad* terminates here.

Toledo is the nearest point for the immense country traversed by these canals and railroads, where a transfer can be made of freight to the more cheap transportation by the lakes, and thence through the Erie Canal, Welland Canal, or Oswego Canal, to the seaboard. It is not merely the country traversed by these canals and railroads that send their products, and receive their merchandise, through Toledo, but many portions of the States of Kentucky, Tennessee, and Missouri, find Toledo the cheapest and most expeditious lake-port for the interchange and transfer of their products and merchandise.

This city is the capital of Lucas County, Ohio, where is situated a court-house and jail, several fine churches, a magnificent High School edifice, and five large brick ward school houses; a young men's association that sustains a course of lectures during the winter; two banks, two insurance companies, six hotels, and a great number of stores and storehouses; also several extensive manufacturing establishments. The principal hotels are the *Island House* and *Oliver House*.

The population of Toledo in 1850 was about 4,000, and now it is supposed to contain 17,000 inhabitants, and is rapidly increasing in wealth and numbers. The shipping interest is increasing, here being trans-shipped annually an amount of grain exceeded only by Chicago, and other kinds of agricultural products of the great West. This city is destined, like Chicago, to export direct to European ports.

At this time there are in process of erection in Toledo many handsome dwellings, numerous handsome blocks of stores, a post-office and custom-house by the general government, and a first-class hotel; these two latter buildings, from the plans we have seen, would do credit to any city, and when completed can be classed among the most elegant structures. No city in the State can boast of finer private residences than Toledo; and the general character of the buildings erected in the past four years is substantial and elegant.

PERRYSBURGH, the capital of Wood Co., Ohio, is situated on the right bank of the Maumee River, 18 miles above its entrance into Maumee Bay, the southern termination of Lake Erie. It contains a court-house and jail, four churches, 20 stores of different kinds, three steam saw-mills, a tannery, and several other manufacturing establishments. Population about 1,500. Here is the head of steamboat navigation on the Maumee River, affording thus far a sufficient depth of water for steamers of a large class.

Old *Fort Meigs*, famous for having withstood a siege by the British and Indians in 1813, is one mile above this place.

MAUMEE CITY, Lucas Co., Ohio, is a port of entry, situated on the Maumee River, opposite Perrysburgh, at the foot of the rapids and at the head of navigation, nine miles above Toledo. A side cut here connects the *Wabash and Erie Canal* with the river. The Toledo and Illinois Railroad also passes through this place. It contains five churches, ten stores, four flouring-mills, three saw-mills, one oil-mill, and other manufacturing establishments propelled by water-power, the supply being here almost inexhaustible. Three miles above the city is the site of the famous battle fought against the Indians by Gen. Wayne, in 1794, known as the Battle of Miami Rapids. One mile below the town is Old *Fort Miami*, one of the early British posts.

MAUMEE RIVER rises in the northeast part of Indiana, and flowing northeast enters Lake Erie, through *Maumee Bay.* It is about 100 miles long, navigable 18 miles, and furnishing an extensive water-power throughout its course.

The City of MONROE, capital of Monroe Co., Mich., is situated on both sides of the River Raisin, three miles above its entrance into Lake Erie, and about 40 miles from Detroit. It is connected with the lake by a ship canal, and is a terminus of the *Michigan Southern Railroad*, which extends west, in connection with the Northern Indiana Railroad, to Chicago, Ill. The town contains about 4,000 inhabitants, a court-house and jail, a United States land-office, eight churches, several public houses, and a number of large stores of different kinds. Here are two extensive piers, forming an outport at the mouth of the river; the railroad track running to the landing. A plank-road also runs from the outport to the city, which is an old and interesting locality, being formerly called *Frenchtown*, which was known as the scene of the battle and massacre of River Raisin in the war of 1812. The *Detroit, Monroe, and Toledo Railroad*, just completed, passes through this city. Steamers run from Detroit to Toledo, stopping at Monroe.

TRENTON, situated on the west bank of Detroit river, is a steamboat landing and a place of considerable trade. Population, 1,000.

WYANDOTTE, ten miles below Detroit, is a new and flourishing manufacturing village, where are located the most extensive Iron Works in Michigan. The iron used at this establishment comes mostly from Lake Superior, and is considered equal in quality to any in the world. The village contains about 1,600 inhabitants.

Railroad Route around Lake Erie.

This important body of water being encompassed by a band of iron, we subjoin the following *Table of Distances*:

	Miles.
Buffalo to Paris, C. W., via *Buffalo and Lake Huron Railroad*,	84
Paris to Windsor or Detroit, via *Great Western Railway*,	158
Detroit to Toledo, Ohio, via *Detroit and Toledo R. R.*,	63
Toledo to Cleveland, via *Cleveland and Toledo R. R.*,	107
Cleveland to Erie, Pa., via *Cleveland and Erie R. R.*,	95
Erie to Buffalo, via *Lake Shore Road*,	88
Total miles,	595

The extreme length of Lake Erie is 250 miles, from the mouth of Niagara River to Maumee Bay; the circuit of the lake about 560 miles, being about 100 miles less distance than has been stated by some writers on the great lakes.

Ohio River and Lake Erie Canals.

The completion of the MIAMI CANAL makes four distinct channels of communication from the Ohio River through the State of Ohio to Lake Erie, namely:
1. The *Erie Extension Canal*, from Beaver, twenty or thirty miles below Pittsburgh, to Erie, 136 miles. 2. The *Cross-Cut Beaver Canal*, which is an extension or branch from Newcastle, Pa., on the Beaver Canal, to Akron, Ohio, where it unites with the Portsmouth and Cleveland Canal—making a canal route from Beaver to Cleveland of 143 miles. 3. The *Ohio Canal*, from Cleveland to Portsmouth, through the centre of the State, 309 miles. 4. The *Miami Extension*, which is a union of the Miami Canal with the Wabash and Erie Canal, through Dayton, terminating at Toledo, at the mouth of the Maumee River on Lake Erie, 247 miles. The vast and increasing business of the Ohio Valley may furnish business for all these canals. They embrace rich portions of Pennsylvania, Ohio, and Indiana; but are not so located as to be free from competition with one another. At no distant time, they would unquestionably command a sufficient independent business, were it not probable that they may be superseded by railways. The capacity of railways—both for rapid and cheap transportation—as it is developed by circumstances and the progress of science, is destined to affect very materially the value and importance of canals.

Fort Wayne.
The United States government is now engaged in making extensive improvements at *Fort Wayne*, which, when completed, will render it one of the strongest fortifications in the country, and almost impregnable against a land assault. The site of the fort, as is well known, is in Springwells, about three miles below the city of Detroit. Its location is admirable, being on a slight eminence, completely commanding the river, which at that point is narrower than in any other place of its entire length. Guns properly placed there could effectually blockade the river against ordinary vessels, and, with the aid of a few gunboats, could repulse any fleet which might present itself.

The present works were erected about the years 1842-'43, mainly under the supervision of General Meigs. The form of the works is that of a star, mounting thirty-two barbette guns at the angles which rake the moat, and protect it against an assault by land. The height from the bottom of the ditch is about forty feet. The exterior of the embankments was supported by timbers, which, in the twenty years in which they have stood, have become unsound, and now give unmistakable evidence of decay. This fact has rendered necessary the improvements which are now being made. They consist of a wall around the entire fort, built against these timbers, which will not be removed, and which will not only sustain the embankments, but will render the place much more impregnable. The wall is seven feet and a half in thickness, and twenty feet in height on every side. The outside facing, two feet in thickness, is of brick, the remainder is filled in with pounded stone, water-lime, sand, and mortar, making a solid wall of great strength. From the nature of the surrounding grounds, artillery cannot be brought to bear upon the wall, with the exception of about two feet at the top, which extends above the level of the ditch. The wall, therefore, could not be battered down, and the only possible way by which the place could be taken would be by a land assault and scaling the walls from the moat by means of ladders. This is effectually provided against by the placing of the guns, eight of which rake the moat on each side. The improvements now being made still further contemplate placing these guns in casemates, which will render them still more secure, protecting the guns and gunners. Barbette guns will also be mounted on the bastions in addition to the casemate guns, which will be placed in a manner similar to that in which they are now placed. These improvements will greatly strengthen the works and render them more permanent. The place is an important one, as the expense the government is at in rendering it impregnable clearly shows. In case of a war with Great Britain it would become of the highest importance. And acting on the maxim, "in time of peace prepare for war," it is the best time to attend to these improvements.

TABLE OF DISTANCES

FROM CLEVELAND AND DETROIT TO SUPERIOR CITY, FORMING A GRAND STEAMBOAT EXCURSION OF OVER TWO THOUSAND MILES.

Ports, &c.	Place to Place.	Miles.	Ports, &c.	Place to Place.	Miles.
CLEVELAND, Ohio	0	0	SUPERIOR City, Wis.	0	0
Malden, C. W	100	100	Point de Tour	70	70
DETROIT, Mich	20	120	Bayfield, Wis.	10	80
Lake St. Clair	7	127	La Pointe "	4	84
Algonac, Mich	33	160	Ontonagon, Mich	74	158
Newport, "	6	166	Eagle River "	60	218
St. Clair, "	10	176	Eagle Harbor "	10	228
Port Huron " / Port Sarnia, C. W.	17	193	Copper Harbor "	16	244
			Manitou Island	15	259
Point au Barque— / Off Saginaw Bay	67	260	Portage Entry	55	314
			Houghton, Mich	(on Portage Lake).	
Thunder Bay Island	75	335	Marquette "	70	384
Point de Tour / St. Mary's River	85	420	Grand Island	40	424
			Pictured Rocks	10	434
Church's Landing	36	456	Point au Sable	20	454
SAUT STE. MARIE	14	470	White Fish Point	50	504
Point Iroquois	15	485	Point Iroquois	25	529
White Fish Point	25	510	SAUT STE. MARIE	15	544
Point au Sable	50	560	Church's Landing	14	558
Pictured Rocks	20	580	Point de Tour, / Lake Huron,	86	594
Grand Island	10	590			
Marquette, Mich	40	630	Thunder Bay Island	85	679
Portage Entry	70	700	Off Saginaw Bay / Point au Barque	75	754
Houghton, Mich	(on Portage Lake).				
Manitou Island	55	755	Port Huron, Mich. / Port Sarnia. C. W.	67	821
Copper Harbor	15	770			
Eagle Harbor	16	786	St. Clair, Mich.	17	838
Eagle River	10	796	Newport "	10	848
Ontonagon, Mich	60	856	Algonac, Mich	6	854
La Pointe, Wis.	74	930	St. Clair Flats	10	864
Bayfield, Wis.	4	934	DETROIT, Mich	30	894
Point de Tour	10	944	Malden, C. W.	20	914
SUPERIOR, City, Wis.	70	1,014	CLEVELAND, Ohio	100	1,014

Detroit, "THE CITY OF THE STRAITS," a port of entry, and the great commercial mart of the State, is favorably situated in N. lat. 42° 20', W. long. 82° 58', on a river or strait of the same name, elevated some 30 or 40 feet above its surface, being seven miles below the outlet of Lake St. Clair and twenty above the mouth of the river, where it enters into Lake Erie. It extends for the distance of upward of a mile upon the southwest bank of the river, where the stream is three-fourths of a mile in width. The principal public and private offices and wholesale stores are located on Jefferson and Woodward avenues, which cross each other at right angles, the latter running to the water's edge. There may usually be seen a great number of steamboats, propellers, and sail vessels of a large class, loading or unloading their rich cargoes, destined for Eastern markets or for the *Great West*, giving an animated appearance to this place, which is aptly called the *City of the Straits*. It was incorporated in 1815, being now divided into ten wards, and governed by a mayor, recorder, and board of aldermen. Detroit contains the old State-house, from the dome of which a fine view is obtained of the city and vicinity; the City Hall. Masonic Hall, Firemen's Hall, Mechanic's Hall, Odd Fellows' Hall, the Young Men's Society Building. two Market Buildings, forty Churches, ten Hotels, besides a number of taverns; a United States Custom-house and Post-office, and United States Lake Survey office, a theatre, a museum, two orphan asylums, four banks, and a savings' fund, institute, water-works, and gas-works, four grain elevators, five steam grist-mills, and several steam saw-mills, besides a great number of other manufacturing establishments. There are also several extensive ship-yards and machine-shops, where are built and repaired vessels of almost every description. The population in 1850 was 21,891; in 1860, 45,619.

The principal Hotels are the *Biddle House*, and *Michigan Exchange*, on Jefferson avenue, and the *Russell House*, on Woodward avenue, facing *Campus Martius*, an open square near the centre of the City.

Detroit may be regarded as one of the most favored of all the Western cities of the Union. It was first settled by the French explorers as early as 1701, as a military and fur trading port. It changed its garrison and military government in 1760 for a British military commander and troops, enduring under the latter *régime* a series of Indian sieges, assaults, and petty but vigilant and harassing warfare, conducted against the English garrison by the celebrated Indian warrior Pontiac. Detroit subsequently passed into possession of the American revolutionists; but on the 16th August, 1812, it was surrendered by Gen. Hull, of the United States army, to Gen. Brock, commander of the British forces. In 1813 it was again surrendered to the Americans, under Gen. Harrison.

The following Railroad lines diverge from Detroit:

1. The *Detroit, Monroe, and Toledo Railroad*, 62 miles in length, connecting with the Michigan Southern Railroad at Monroe, and with other roads at Toledo.

2. The *Michigan Central Railroad*, 282 miles in length, extends to Chicago, Ill. This important road, running across the State from east to west, connects at

Michigan City, Ind., with the New Albany and Salem Railroad—thus forming a direct line of travel to Louisville, St. Louis, etc., as well as Chicago and the Far West.

3. The *Detroit and Milwaukee Railroad* runs through a rich section of country to Grand Haven, on Lake Michigan, opposite Milwaukee, Wis.

4. The *Detroit and Port Huron Railroad*, connecting with the Grand Trunk Railway of Canada, connects Lake Huron by rail with the valley of the Ohio River.

5. The *Great Western Railway* of Canada has its terminus at Windsor, opposite Detroit, the two places being connected by three steam ferries—thus affording a speedy line of travel through Canada, and thence to Eastern cities of the United States.

Steamers of a large class run from Detroit to Cleveland, Toledo, and other ports on Lake Erie ; others run to Port Huron, Saginaw, Goodrich, C. W., and other ports on Lake Huron.

The *Lake Superior* line of steamers running from Cleveland and Detroit direct for the Saut Ste. Marie, and all the principal ports on Lake Superior, are of a large class, carrying passengers and freight. This has become one of the most fashionable and healthy excursions on the continent.

The DETROIT RIVER, or *Strait*, is a noble stream, through which flow the surplus waters of the Upper Lakes into Lake Erie. It is 27 miles in length, and from half a mile to two miles in width, forming the boundary between the United States and Canada. It has a perceptible current. and is navigable for vessels of the largest class. Large quantities of fish are annually taken in the river, and the sportsman usually finds an abundance of wild ducks, which breed in great numbers in the marshes bordering some of the islands and harbors of the coast.

There are altogether seventeen islands in the river. The names of these are, *Clay, Celeron, Hickory, Sugar, Bois Blanc, Ella, Fox, Rock, Grosse Isle, Stoney, Fighting, Turkey, Mammy Judy, Grassy, Mud, Belle* or *Hog*, and *Ile la Pêche.* The two latter are situated a few miles above Detroit, near the entrance to Lake St. Clair, where large quantities of white-fish are annually taken.

ILE LA PÊCHE, attached to Canada, was the home of the celebrated Indian chief *Pontiac.* Parkman, in his "History of the Conspiracy of Pontiac," says: "Pontiac, the Satan of this forest-paradise, was accustomed to spend the early part of the summer upon a small island at the opening of Lake St. Clair." Another author says: "The king and lord of all this country lived in no royal state. His cabin was a small, oven-shaped structure of bark and rushes. Here he dwelt with his squaws and children; and here, doubtless, he might often have been seen carelessly reclining his naked form on a rush-mat or a bear-skin, like an ordinary Indian warrior."

The other fifteen islands, most of them small, are situated below Detroit, within the first twelve miles of the river after entering it from Lake Erie, the largest of which is GROSSE ISLE, attached to Michigan, on which are a number of extensive and well-cultivated farms. This island has become a very popular retreat for citizens of Detroit during the heat of summer, there being here located good public houses for the accommodation of visitors.

Father Hennepin, who was a passenger on the "Griffin," the first vessel that crossed Lake Erie, in 1679, in his description of the scenery along the route says: "The islands are the finest in the world: the strait is finer than Niagara ; the banks are vast meadows, and the prospect is terminated with some hills covered with vineyards, trees bearing good fruit, groves and forests so well disposed that one would think that Nature alone could not have made, without the help of art, so charming a prospect."

COMPARATIVE PURITY OF DETROIT RIVER WATER.

The following Table shows the solid matter in a gallon of water, taken from Lakes and Rivers in different cities:

Albany, Hudson River	6,320
Troy, Mohawk River	7,880
Boston, Cochituate Lake	1,870
New York, Croton River	6,998
Brooklyn, L. I. Ponds	2,867
Philadelphia, Schuylkill R	4,260
Cincinnati, Ohio River	6,746
Lake Ontario	4,160
Detroit, Detroit River	5,722
Cleveland, Lake Erie	5,080
Montreal, St. Lawrence R	5,000

Of the Detroit River water, Prof. Douglass, in his report of the analysis, says: "In estimating the value of your city water, as compared with other cities, due allowance must be made for the fact, that the total solid matter is materially increased by the presence of silica, alumina, and iron, elements that can produce little or no injury; while the chlorides, much the most injurious compounds, are entirely absent. The presence of such large quantities of silica and iron is accounted for by the fact that Lakes Superior and Huron are formed, for the most part, in a basin of ferruginous sandstone and igneous rock."

NOTE.—The purity of the waters of Lake Superior, probably exceeds all other bodies of water on the face of the globe, affording a cool and delightful beverage at all seasons.

Comparative Increase of Lake Cities.

	1840.	1850.	1860.
BUFFALO, New York	18,213	42,261	81,131*
CHICAGO, Ill	4,470	28,269	109,263
CLEVELAND, Ohio	6,071	17,034	36,054†
DETROIT, Mich	9,102	21,019	45,619
ERIE, Penn		5,858	9,419
MILWAUKEE, Wis	1,700	20,061	45,254
OSWEGO, New York		12,205	16,817
RACINE, Wis		5,107	10,000
SANDUSKY, Ohio	1,434	6,008	8,408
TOLEDO, Ohio	1,222	3,829	13,768

* Black Rock annexed. † Ohio City annexed.

DETROIT AND MILWAUKEE RAILROAD AND STEAMSHIP LINE,

CONNECTING WITH THE GREAT LINES OF TRAVEL EAST AND WEST.

On leaving the Railroad Depot at Detroit the line of this road runs in a northwest direction to PONTIAC, 26 miles, passing through a rich section of farming country.

The route then continues westerly to FENTONVILLE, 24 miles further, where commences a railroad route, running through Flint, and extending north to Saginaw, favorably situated on Saginaw river. It is intended to continue the Flint and Pere Marquette Railroad, some 150 miles, to the shore of Lake Michigan.

OWASSO, 78 miles from Detroit, and 110 miles from Grand Haven, is an important station, from whence a railroad extends southwest to LANSING, the capital of the State of Michigan. It is intended to carry the line of this road north to Saginaw City, and from thence northwest to Traverse Bay on Lake Michigan, where is a good harbor.

From Owasso, the Detroit and Milwaukee Railroad runs westward through St. John's, Ionia, and other stations, passing down the valley of the Grand River, a rich and populous section of country, producing large quantities of wheat and other agricultural productions, all of which find a ready sale in the Eastern markets.

5

"Up in the northern part of the Grand River Valley, and along and beyond the Muskegon River, an immense amount of pine timber is to be found, giving profitable employment to a large number of lumbermen."

Grand Rapids, 158 miles west of Detroit, and forty miles above Grand Haven, an incorporated city, is favorably situated on both banks of Grand River, where is a fall of about eighteen feet, affording an immense water-power. Steamers run from this place daily to Grand Haven, connecting with steamers for Milwaukee, Chicago, and other ports on Lake Michigan. Here is an active population of about 10,000, and rapidly increasing, surrounded by a new, fertile, and improving country, being alike famous as a wheat and fruit region.

The city now contains a court-house and jail; 6 churches; 5 hotels; 60 stores of different kinds; 3 grist-mills; 5 saw-mills; 3 cabinet-ware factories; 2 machine-shops, and other manufacturing establishments. The private dwellings and many of the stores are elegant edifices, constructed of building material which is found in the immediate vicinity. Extensive and inexhaustible beds of gypsum are found near this place, producing large quantities of stucco and plaster, all of which find a ready sale in Eastern and Western markets. It is estimated that 25,000 tons of stucco for building, and plaster for fertilizing purposes, can be quarried and ground yearly from the different quarries in this vicinity. At the Eagle Mills, two miles below the city, is already formed an immense excavation, extending several hundred feet under ground, which is well worthy of a visit, where rich specimens of the gypsum can be obtained.

Steamboat Route from Grand Rapids to Grand Haven.

GRAND RAPIDS.		0
Eagle Plaster Mill.		2
Grandville.	5	7
Lamont.	13	20
Eastmanville.	2	22
Mill Point.	16	38
GRAND HAVEN.	2	40

On leaving Grand Rapids for Grand Haven, by railroad, the route extends north of the river, through a fertile section of country, mostly covered by a heavy growth of hardwood, although the pine predominates as you approach the lake shore.

Grand Haven, Ottawa Co., Mich., is situated on both sides of Ottawa River, near its entrance into Lake Michigan, here eighty-five miles wide; on the opposite side lies Milwaukee, Wisconsin. The different settlements, comprising Grand Haven, contain about 5,000 inhabitants. Here are a court-house and jail; 3 churches; six hotels and taverns; and a number of stores and warehouses; 8 large steam saw-mills, pail and tub factories, a foundry and machine-shop, and other manufacturing establishments. Steamers and sail vessels run from Grand Haven, which has a well-protected harbor, to Milwaukee, Chicago, and other ports on Lake Michigan, carrying a large amount of produce and lumber. The fisheries in this vicinity are also productive and extensive.

The sand hills on the east shore of Lake Michigan rise from 100 to 200 feet, presenting a sterile appearance, although the land in the interior is very rich and productive.

Trip across Lake Michigan.

The staunch and well-built steamships, *Detroit*, Capt. McBride, and *Milwaukee*, Capt. Trowell, run twice daily across Lake Michigan, connecting with trains on the Detroit and Milwaukee railroad. This trip is delightful during the summer and autumn months when Lake Michigan is usually calm, affording a safe and delightful excursion of about six hours' continuance. The spacious cabins, and well-arranged dining-saloons of these ships, together with the well-provided tables, renders this route to and from the Eastern cities one of the most pleasant and desirable as regards speed and objects of interest. Usual fare from Milwaukee to Detroit, $8,00. Distance, 271 miles; time, 14 hours.

From Milwaukee, westward, there is a direct connection both with the *Milwaukee and Prairie du Chien Railroad*, and the *La Crosse and Milwaukee Railroad* running to the Mississippi River. A daily line of steamers run from the termination of both of the above railroads to St. Paul, Minnesota.

TRIP FROM DETROIT TO MACKINAC, GREEN BAY, MILWAUKEE, CHICAGO, &c.

During the season of navigation propellers of a large class, with good accommodations for passengers, leave Detroit daily direct for Mackinac, Green Bay, Milwaukee, and Chicago, situated on Lake Michigan.

Steamers of a large class, carrying passengers and freight, also leave Detroit, almost daily for the Saut Ste Marie, from thence passing through the *Ship Canal* into Lake Superior—forming delightful excursions during the summer and early autumn months.

For further information of steamboat routes, see *Advertisements*.

On leaving Detroit the steamers run in a northerly direction, passing *Bell* or *Hog Island*, two miles distant, which is about three miles long and one mile broad, presenting a handsome appearance. The Canadian shore on the right is studded with dwellings and well cultivated farms.

PECHE ISLAND is a small body of land attached to Canada, lying at the mouth of Detroit River, opposite which, on the Michigan shore, is *Wind-Mill Point* and light-house.

LAKE ST. CLAIR commences seven miles above Detroit; it may be said to be 20 miles long and 25 miles wide, measuring its length from the outlet of St. Clair River to the head of Detroit River. Compared with the other lakes it is very shallow, having a depth of only from 8 to 24 feet as indicated by Bayfield's chart. It receives the waters of the Upper Lakes from the St. Clair Strait by several channels forming islands, and discharges them into the Detroit River or Strait. In the upper portion of the lake are several extensive islands, the largest of which is *Walpole Island*; it belongs to Canada, and is inhabited mostly by Indians. All the islands to the west of Walpole Island belong to Michigan. The Walpole, or "Old Ship Channel," forms the boundary between the United States and Canada. The main channel, now used by the larger class of vessels, is called the "North Channel." Here are passed the "*St Clair Flats*," a great impediment to navigation, for the removal of which Congress will no doubt make ample appropriation sooner or later. The northeastern channel, separating Walpole Island from the main Canada shore, is called "*Chenail Ecarte.*" Besides the waters passing through the Strait of St. Clair, Lake St. Clair receives the river Thames from the Canada side, which is navigable to Chatham, some 24 miles; also the waters of Clinton River from the west or American side, the latter being navigable to Mt. Clemens, Michigan. Several other streams flow into the lake from Canada, the principal of which is the River Sydenham. Much of the land bordering on the lake is low and marshy, as well as the islands; and in places there are large plains which are used for grazing cattle.

ASHLEY, or NEW BALTIMORE, situated on the N. W. side of Lake St. Clair, 30 miles from Detroit, is a new and flourishing place, and has a fine section of country in the rear. It contains three steam saw-mills, several other manufactories, and about 1,000 inhabitants. A steamboat runs from this place to Detroit.

MT. CLEMENS, Macomb Co., Mich., is situated on Clinton River, six miles above its entrance into Lake St. Clair, and about 30 miles from Detroit by lake and river. A steamer plies daily to and from Detroit during the season of navigation. Mt. Clemens contains the county buildings, several churches, three hotels, and a number of stores and manufacturing establishments, and about 2,000 inhabitants. Detroit is distant by plank road only 20 miles.

CHATHAM, C. W., 46 miles from Detroit by railroad route, and about 24 miles above the mouth of the river Thames, which enters into Lake St. Clair, is a port of entry and thriving place of business, where have been built a large number of steamers and sail-vessels.

ALGONAC, Mich., situated near the foot of St. Clair River, 40 miles from Detroit, contains a church, two or three saw-mills, a grist-mill, woollen factory, and about 700 inhabitants.

NEWPORT, Mich., seven miles farther north, is noted for steamboat building, there being extensive ship-yards, where are annually employed a large number of workmen. Here are four steam saw-mills, machine shops, etc. Population about 1,200. Belle River here enters the St. Clair from the west.

ST. CLAIR STRAIT connects Lake Huron with Lake St. Clair, and discharges the surplus waters of Lakes Superior, Michigan, and Huron. It flows in a southerly direction, and enters Lake St. Clair by six channels, the north one of which, on the Michigan side, is the only one at present navigated by large vessels in ascending and descending the river. It receives several tributaries from the west, or Michigan; the principal of which are Black River, Pine River, and Belle River, and several rivers flow into it from the east, or Canadian side. It has several flourishing villages on its banks. It is 48 miles long, from a half to a mile wide, and has an average depth of from 40 to 60 feet, with a current of three miles an hour, and an entire descent of about 15 feet. Its waters are clear and transparent, the navigation easy, and the scenery varied and beautiful—forming for its entire length, the boundary between the United States and Canada. The banks of the upper portion are high; those of the lower portion are low and in parts inclined to be marshy. Both banks of the river are generally well settled, and many of the farms are beautifully situated. There are several wharves constructed on the Canada side, for the convenience of supplying the numerous steamboats passing and repassing with wood. There is also a settlement of the Chippewa Indians in the township of Sarnia, Canada; the Indians reside in small log or bark houses of their own erection.

The CITY OF ST. CLAIR, Mich., is pleasantly situated on the west side of St. Clair Strait, 56 miles from Detroit and 14 miles from Lake Huron. This is a thriving place, with many fine buildings, and is a great lumber depôt. It contains the county buildings for St. Clair Co., several churches and hotels, one flouring-mill, and five steam saw-mills, besides other manufacturing establishments, and about 3,000 inhabitants. St. Clair has an active business in the construction of steamers and other lake craft. The site of old *Fort St. Clair*, now in ruins, is on the border of the town.

SOUTHERLAND, C. W., is a small village on the Canada shore, opposite St. Clair. It was laid out in 1833 by a Scotch gentleman of the same name, who here erected an Episcopal church, and made other valuable improvements.

MOORE, is a small village ten miles below Sarnia on the Canada side.

FROMEFIELD, or TALFOURD'S, C. W., is another small village, handsomely situated four and a half miles below Sarnia. Here is an Episcopal church, a windmill, and a cluster of dwellings.

The city of PORT HURON, St. Clair Co., Mich., a port of entry, is advantageously situated on the west bank of St. Clair River, at the mouth of Black River, two miles below Lake Huron. It was chartered in 1858, and now contains one Congregational, one Episcopal, one Baptist, one Methodist, and one Roman Catholic Church; six hotels, and public houses, forty stores, and several warehouses; one steam flouring-mill, eight steam saw-mills, producing annually a large amount of lumber, the logs being rafted down Black River, running through an extensive pine region; here are also, two yards for building of lake craft, two refineries of petroleum oil, one iron foundry, and several other manufacturing establishments. Population in 1860, 4,000.

During the season of navigation, there is daily intercourse by steamboats with Detroit, Saginaw, and ports on the Upper Lakes. A steam ferry-boat also plies between Port Huron and Sarnia, C. W., the

St. Clair River here being about one mile in width. A branch of the Grand Trunk Railway runs from Fort Gratiot, one mile and a half above Port Huron, to Detroit, a distance of 62 miles, affording altogether speedy modes of conveyance. A railroad is also proposed to run from Port Huron, to intersect with the Detroit and Milwaukee Railroad, at Owasso, Michigan.

FORT GRATIOT, one and a half miles north of Port Huron, lies directly opposite Point Henry, C. W., both situated at the foot of Lake Huron, where commences St. Clair River. It has become an important point since the completion of the Grand Trunk Railway of Canada, finished in 1859, which road terminates by a branch at Detroit, Mich., thus forming a direct railroad communication from Lake Huron, eastward, to Montreal, Quebec, and Portland, Maine.

The village stands contiguous to the site of Fort Gratiot, and contains besides the railroad buildings, which are extensive, one church, five public houses, the Gratiot House being a well-kept hotel; two stores, one oil refinery, and about 400 inhabitants. A steam ferry-boat plies across the St. Clair River, to accommodate passengers and freight; the river here being about 1,000 feet wide, and running with considerable velocity, having a depth of from 20 to 60 feet.

In a military and commercial point of view, this place attracts great attention, no doubt, being destined to increase in population and importance. The Fort was built in 1814, at the close of the war with Great Britain, and consists of a stockade, including a magazine, barracks, and other accommodations for a garrison of one battalion. It fully commands the entrance to Lake Huron from the American shore, and is an interesting landmark to the mariner.

SARNIA, C. W., situated on the east bank of St. Clair River, two miles below

Lake Huron and 68 above Detroit, is a port of entry and a place of considerable trade; two lines of railroad terminate at this point, and it is closely connected with Port Huron on the American shore by means of a steam ferry. The town contains a court-house and jail, county register's office and town hall; one Episcopal, one Methodist, one Congregational, one Baptist, one Roman Catholic, and one Free Church; seven public houses, the principal being the *Alexander House* and the *Western Hotel;* twenty stores and several groceries; two grain elevators, two steam saw-mills; one steam grist-mill, one large barrel factory, one steam cabinet factory, one steam iron foundry, and one refinery of petroleum oil, besides other manufacturing establishments. Population, 2,000.

The Grand Trunk Railway of Canada terminates at Point Edward, 2 miles from Sarnia, extending eastward to Montreal, Quebec, and Portland, Me.; a branch of the Great Western Railway also terminates at Sarnia, affording a direct communication with Niagara Falls, Boston, and New York. Steamers run from Sarnia to Goderich and Saugeen, C. W.; also to and from Detroit, and ports on the Upper Lakes.

The celebrated *Enniskillen Oil Wells,* yielding an immense quantity of petroleum oil of a superior quality, are distant some 18 or 20 miles from Sarnia, this being the nearest shipping port. These wells are easy of access by means of railway and plank-roads; the oil is brought to Sarnia in barrels, and much of it shipped from hence direct to European ports, passing down the St. Lawrence River.

The St. Clair River, opposite Sarnia, here one mile in width, flows downward with a strong current, at the rate of about six miles an hour.

Steamboat Route from Sarnia to Goderich, Saugeen, etc.

Steamers running to and from Detroit on their way to the different ports on the east shore of Lake Huron, usually hug the Canada side, leaving the broad waters of the lake to the westward.

POINT EDWARD, 2 miles above Sarnia, lies at the foot of Lake Huron, opposite Fort Gratiot, where are erected a large depôt and warehouses connected with the *Grand Trunk Railway* of Canada. Here terminates the grand railroad connection extending from the Atlantic ocean to the Upper Lakes. It also commands the entrance into Lake Huron and is an important military position although at present unfortified. In the vicinity is an excellent fishery, from whence large quantities of fish are annually exported.

BAYFIELD, C. W., 108 miles from Detroit, is a new and flourishing place, situated at the mouth of a river of the same name.

GODERICH, 120 miles north of Detroit, is situated on elevated ground at the mouth of Maitland River, where is a good harbor. This is a very important and growing place, where terminates the *Buffalo and Huron Railroad,* 160 miles in length. (*See* page 53.)

KINCARDINE, thirty miles from Goderich, is another port on the Canadian side of Lake Huron, where the British steamers land and receive passengers on their trips to Saugeen.

SAUGEEN, C. W., is situated at the mouth of a river of the same name, where is a good harbor for steamers and lake craft. This is the most northern port to which steamers now run on the Canada side of Lake Huron, and will no doubt, ere long be reached by railroad.

Steamboat Route from Port Huron to Saginaw City, etc.

On leaving the wharf at Port Huron, the steamers pass Fort Gratiot and enter the broad waters of Lake Huron, one of the Great Upper Lakes, all alike celebrated for the sparkling purity of their waters. The shores are for the most part low, being covered by a heavy growth of forest trees.

LAKEPORT, 11 miles from Port Huron, is a small village lying on the lake shore.

LEXINGTON, 11 miles further, is the capital of Sanilac County, Michigan, where is a good steamboat landing and a flourishing settlement.

PORT SANILAC, 34 miles above Port Huron, is another small settlement.

FORRESTVILLE, Mich., 47 miles from Port Huron, and 120 miles north of Detroit, situated on the west side of Lake Huron, is a new settlement, where is erected an extensive steam saw-mill. It has some three or four hundred inhabitants, mostly engaged in the lumber trade. Several other small settlements are situated on the west shore of Lake Huron, which can be seen from the ascending steamer, before reaching Point aux Barques, about seventy-five miles above Port Huron.

SAGINAW BAY is next entered, presenting a wide expanse of waters; Lake Huron here attaining its greatest width, where the mariner often encounters fierce storms, which are prevalent on all of the Upper Lakes. To the eastward lies the Georgian Bay of Canada, with its innumerable islands.

BAY CITY, or LOWER SAGINAW, near the mouth of Saginaw River, is a flourishing town, with a population of about 2,500. Here is a good harbor, from whence a large amount of lumber is annually exported. It has fifteen saw-mills, and other manufacturing establishments.

Steamers run daily to Detroit and other ports.

EAST SAGINAW, situated on the right bank of the river, about one mile below Saginaw City, is a new and flourishing place, and bids fair to be one of the most important cities of the state. It is largely engaged in the lumber trade, and in the manufacture of salt of a superior quality. There are several large steam saw-mills, many with gangs of saws, and capable of sawing from four to five million feet of lumber annually; grist and flouring-mills, with four run of stones, planing-mills, foundries, machine shops, breweries, a ship-yard, and other manufacturing establishments, giving employment to a great number of workmen. Here is a well-kept hotel, and several churches; a banking office and a number of large stores and warehouses. Coal of a good quality is abundant, being found near the river, and the recent discovery of salt springs in the neighborhood is of incalculable value, the manufacture of salt being carried on very extensively. Population, 4,500.

Several lines of steamers, and one of propellers, sail from this port regularly for Detroit and other lake ports. It is near the head of navigation for lake craft, where five rivers unite with the Saginaw, giving several hundred miles of water communication for river rafting and the floating of saw-logs. The surrounding country is rich in pine, oak, cherry, blackwalnut, and other valuable timber. A railroad is finished from this place to *Flint*, connecting by stages with the Detroit and Milwaukee railroad.

SAGINAW CITY, Saginaw County, Mich., is handsomely situated on the left bank of the river, 23 miles above its mouth. It contains a court-house and jail, several churches, two hotels, fifteen stores, two warehouses, and six steam saw-mills. Population about 3,000. There is a fine section of country in the rear of Saginaw, much of which is heavily timbered; the soil produces grain in abundance, while the streams afford means of easy transportation to market. Steamers run daily from Saginaw City and East Saginaw to Detroit, Chicago, &c., and other ports on the lakes, during the season of navigation.

LAKE HURON.

The waters of Lake Huron, lying between 43° and 46° north latitude, are surrounded by low shores on every side. The most prominent features are Saginaw Bay on the southwest, and the Georgian Bay on the northeast; the latter large body of water being entirely in the limits of Canada. The lake proper, may be said to be 100 miles in width, from east to west, and 250 miles in length, from south to north, terminating at the Straits of Mackinac. It is nearly destitute of islands, presenting one broad expanse of waters. It possesses several good harbors on its western shores, although as yet but little frequented. Point aux Barques, Thunder Bay, and Thunder Bay Islands, are prominent points to the mariner.

TAWAS, or OTTAWA BAY, lying on the northwest side of Saginaw Bay, affords a good harbor and refuge during storms, as well as THUNDER BAY, lying farther to the north. Off Saginaw Bay, the widest part of the lake, rough weather is often experienced, rendering it necessary for steamers and sail vessels to run for a harbor or place of safety.

In addition to the surplus waters which Lake Huron receives through the Straits of Mackinac and the St. Mary's River from the north, it receives the waters of Saginaw River, and several other small streams from the west. This lake drains but a very small section of country compared to its magnitude, while its depth is a matter of astonishment, being from 100 to 750 feet, according to recent surveys;

altitude above the ocean, 574 feet, being 26 feet below the surface of Lake Superior. Its outlet, the St. Clair River, does not seem to be much larger than the St. Mary's River, its principal inlet, thus leaving nearly all its other waters falling in the *basin*, to pass off by evaporation. On entering the *St. Clair River*, at Fort Gratiot, after passing over the Upper Lakes, the beholder is surprised to find all these accumulated waters compressed down to a width of about 1,000 feet, the depth varying from 20 to 60 feet, with a strong downward current.

The *Straits of Mackinac*, connecting Lakes Huron and Michigan, is a highly interesting body of water, embosoming several picturesque islands, with beautiful headlands along its shores. It varies in width from 5 to 30 miles, from mainland to mainland, and may be said to be from 30 to 40 miles in length. Here are good fishing grounds, as well as at several other points on Lake Huron and Georgian Bay.

The climate of Lake Huron and its shores is perceptibly warmer than Lake Superior during the spring, summer, and autumn months, while the winter season is usually rendered extremely cold from the prevalence of northerly winds passing over its exposed surface. On the 30th of July, 1860, at 8 A. M., the temperature of the air near the middle of Lake Huron, was 64° Fahr., the water on the surface, 52°, and at the bottom, 50 fathoms (300 feet) 42° Fahr.

THE LOWER PENINSULA OF MICHIGAN.

THE *Lower Peninsula of Michigan* is nearly surrounded by the waters of the Great Lakes, and, in this respect, its situation is naturally more favorable for all the purposes of trade and commerce than any other of the Western States.

The numerous streams which penetrate every portion of the Peninsula, some of which are navigable for steamboats a considerable distance from the lake, being natural outlets for the products of the interior, render this whole region desirable for purposes of settlement and cultivation. Even as far north as the Strait of Mackinac, the soil and climate, together with the valuable timber, offer great inducements to settlers; and if the proposed railroads, under the recent grant of large portions of these lands by Congress, are constructed from and to the different points indicated, this extensive and heavily timbered region will speedily be reclaimed, and become one of the most substantial and prosperous agricultural portions of the West.

It is well that in the system of compensation, which seems to be a great law of the universe, the vast prairies which comprise so large a portion of this great Western domain are provided so well with corresponding regions of timber, affording the necessary supply of lumber for the demand of the increasing population which is so rapidly pouring into these Western States.

The State of Michigan—all the waters of which flow into the Basin of the St. Lawrence—Northern Wisconsin, and Minnesota are the sources from which the States of Ohio, Indiana, Illinois, and Iowa, and a large portion of the prairie country west of the Mississippi, must derive their supply of this important article (lumber). The supply in the West is now equal to the demand, but the consumption is so great, and the demand so constantly increasing with the development and settlement of the country, that of necessity, within comparatively a very few years, these vast forests will be exhausted. But as the timber is exhausted the soil is prepared for cultivation, and a large portion of the northern part of the southern Peninsula of Michigan will be settled and cultivated, as it is the most reliable wheat-growing portion of the Union.

Besides the ports and towns already described, there are on *Lake Huron*, after leaving *Saginaw Bay*, going north, several settlements and lumber establishments, fisheries, &c. These are at *Tawas Bay*, mouth of the River au Sable, Black River, &c.

ALPENA, situated at the head of Thunder Bay, is a very flourishing town, and the capital of Alpena County. It contains about 500 inhabitants, and four saw-mills, possessing a superior water-power on the river here emptying into the bay. It is both a lumber and fishing station of considerable importance.

DUNCAN is the next place of importance on the lake coast, situated near the mouth of Cheboygan River. The United States Land Office for this district is located at this place. Nearly opposite lies *Bois Blanc Island*, a large and fertile tract of land.

The celebrated ISLAND OF MACKINAC is next reached, lying within the straits, surrounded by a cluster of interesting points of land justly celebrated in Indian legends and traditions.

OLD MACKINAC, lying on the mainland, is one of the most interesting points, being celebrated both in French and English history when those two great powers contended for the possession of this vast Lake Region. It is proposed to build a railroad from Old Mackinac to Saginaw, and one to the southern confines of the State, while another line of road will extend northwestward to Lake Superior, crossing the straits by a steam ferry. A town plot has been surveyed, and preparations made for settlement.

Passing around the western extremity of the Peninsula, at the *Waugoshance* Light and Island, the next point is *Little Traverse Bay*, a most beautiful sheet of water.

About fifteen miles southwesterly from Little Traverse we enter GRAND TRAVERSE BAY, a large and beautiful arm of the lake, extending about thirty miles inland. This bay is divided into two parts by a point of land, from two to four miles wide, extending from the head of the bay about eighteen miles toward the lake. The country around this bay is exceedingly picturesque, and embraces one of the finest agricultural portions of the State. The climate is mild, and fruit and grain of all kinds suitable to a northern latitude are produced, with less liability to injury from frost than in some of the southern portions of the State.

GRAND TRAVERSE CITY is located at the head of the west arm of the bay, and is the terminus of the proposed railroad from Grand Rapids, a distance of about 140 miles.

Passing out of the bay and around the point dividing the west arm from the lake, we first arrive at the river *Aux Becs Sceis.* There is here a natural harbor, capable of accommodating the larger class of vessels and steamboats. A town named FRANKFORT has been commenced at this place, and with its natural advantages, and the enterprise of parties who now contemplate making further improvements, it will soon become a very desirable and convenient point for the accommodation of navigators.

The islands comprising the Beavers, the Manitous, and Fox Isles should here be noticed. The *Beavers* lie a little south of west from the entrance to the Strait of Mackinac, the Manitous a little south of these, and the Foxes still farther down the lake. These are all valuable for fishing purposes, and for wood and lumber. Lying in the route of all the steamboat lines from

Chicago to Buffalo and the Upper Lakes, the harbors on these islands are stopping-points for the boats, and a profitable trade is conducted in furnishing the necessary supplies of wood, etc.

We next arrive at MANISTEE, a small but important settlement at the mouth of the Manistee River. The harbor is a natural one, but requires some improvement. A large trade is carried on with Chicago in lumber.

The next point of importance is the mouth of the *Père Marquette* River. Here is the terminus of the proposed railroad from Flint, in Genesee County, connecting with Detroit by the Detroit and Milwaukee Railway, a distance of about 180 miles.

The harbor is very superior, and the country in the vicinity is well adapted for settlement. About 16 miles in the interior is situated one of the most compact and extensive tracts of pine timber on the western coast.

About forty miles south of this, in the county of Oceana, a small village is located at the mouth of *White River*. The harbor here is also a natural one, and the region is settled to considerable extent by farmers. Lumber is, however, the principal commodity, and the trade is principally with the Chicago market.

The next point, MUSKEGON, at the mouth of the *Muskegon River*, is supported principally by the large lumber region of the interior. Numerous steam saw-mills are now in active operation here, giving the place an air of life and activity.

The harbor is one of the best on the lake, and is at present accessible for all the vessels trading between Muskegon and Chicago.

GRAND HAVEN, Ottawa Co., Mich, is situated on both sides of Grand River, at its entrance into Lake Michigan, here eighty-five miles wide; on the opposite side lies Milwaukee, Wis. *For further description, see page 66.*

DIRECT STEAMBOAT ROUTE FROM DETROIT TO GREEN BAY, CHICAGO, &c.

Sailing direct through Lake Huron to Mackinac, or to the De Tour entrance to St. Mary's River, a distance of about 330 miles, the steamer often runs out of sight of land on crossing Saginaw Bay.

Thunder Bay Light is first sighted and passed, and then Presque Isle Light, when the lake narrows and the Strait of Mackinac is soon entered, where lies the romantic Island of Mackinac. The Strait of Mackinac, with the approaches thereto from Lakes Huron and Michigan, will always command attention from the passing traveller. Through this channel will pass, for ages to come, a great current of commerce, and its shores will be enlivened with civilized life.

In this great commercial route, Lake Huron is traversed its entire length, often affording the traveller a taste of sea-sickness and its consequent evils. Yet there often are times when Lake Huron is hardly ruffled, and the timid passenger enjoys the voyage with as much zest as the more experienced mariner.

MACKINAC, crowned by a fortress, where wave the *Stars and the Stripes*, the gem of the Upper Lake islands, may vie with any other locality for the salubrity of its climate, for its picturesque beauties, and for its vicinity to fine fishing-grounds. Here the invalid, the seeker of pleasure, as well as the sportsman and angler, can find enjoyment to their heart's content during warm weather. *For description, see p. 88.*

On leaving Mackinac for Green Bay, the steamer generally runs a west course for the mouth of the bay, passing the Beaver Islands in Lake Michigan before entering the waters of Green Bay, about 150 miles. SUMMER ISLAND lies on the north side and ROCK ISLAND lies on the south side of the entrance to Green Bay, forming a charming view from the deck of a steamer.

WASHINGTON or POTAWATOMEE ISLAND, CHAMBERS' ISLAND, and other small islands are next passed on the upward trip toward the head of the bay.

WASHINGTON HARBOR, situated at the north end of Washington Island, is a picturesque fishing station, affording a good steamboat-landing and safe anchorage.

GREEN BAY, about 100 miles long and from 20 to 30 miles wide, is a splendid sheet of water, destined no doubt to be enlivened with commerce and pleasure excursions. Here are to be seen a number of picturesque islands and headlands. Several important streams enter into Green Bay, the largest of which is Neenah or Fox River, at its head, and is the outlet of Winnebago Lake. Menomonee River forms the boundary between the States of Wisconsin and Michigan, and empties into the bay opposite Green Island.

The recent improvement of the Fox and Wisconsin Rivers, not only opens steamboat navigation between the Bay and the head of Lake Winnebago, but it connects the Fox and Wisconsin Rivers, one of which, flowing northward, falls into the Atlantic through the St. Lawrence, and the other, running southward, discharges its waters, through the Mississippi, into the Gulf of Mexico. By this connection a steamer can start from New Orleans, pass up the Mississippi to the mouth of the Wisconsin, pass up this river to Portage, through a short canal to the Upper Fox Rivers, down this river to Lake Winnebago, at Oshkosh,—down the lake to the point where it contracts into the Lower Fox,—down this romantic river some thirty-five miles, by means of numerous canals around the principal rapids, into Green Bay, and so on without interruption through the great lakes into the St. Lawrence to the Atlantic Ocean.

Green Bay, one of the most favored cities of Wisconsin, the Capital of Brown County, is advantageously situated near the mouth of Fox or Neenah river, at its entrance into Green Bay, where is a good and secure harbor. It lies 90 miles southwest from Lake Michigan, by water, 25 miles due west of Kewaunee, on the west shore of Lake Michigan, and 115 miles north from Milwaukee. The town is handsomely situated, and contains many large warehouses and elegant residences, together with several churches, hotels, and stores of different kinds, and about 5,000 inhabitants. The improvement of *Fox River* by dams and locks, in connection with the improvements on the Wisconsin River, affords an uninterrupted steam navigation from Green Bay to Prairie du Chien, on the Mississippi River—thus making Green Bay a great point for the trans-shipment of goods and produce of every variety ; the largest class steamers and propellers running to Chicago on the south, Saut Ste. Marie on the north, as well as to Collingwood, to Sarnia, to Detroit, and to Buffalo on the east. The lumber trade of Green Bay is immense, this whole section of country abounding in timber of different kinds the most useful for building purposes. There is no city in the West which can boast of a position so advantageous commercially, or which will compare with it in after years in the wealth and extent of its trade. With an uninterrupted water communication East and South, a harbor five miles

in length, capable of accommodating the shipping of the whole lakes, it will eventually be connected by Railroad with every important point West and North, as the nearest route to and from the Eastern and Southern markets.

ASTOR is the name of a suburb of Green Bay, lying at the mouth of Fox River, while on the opposite side of the stream stands FORT HOWARD, surrounded by a village of the same name, where terminates the *Chicago and Northwest Railway*, running south to Chicago, 242 miles.

OCONTO, situated on Green Bay, at the mouth of the Oconto River, is a new and thriving lumber settlement. It lies 25 miles north of the town of Green Bay, having daily communication by steamboat. As regards the lumbering interests of this region, a late writer says: "*Oconto County* is a portion of the great Pine region of Northern Wisconsin, lying along the west shore of Green Bay, and is, for lumbering purposes, one of the most important counties in the State—being easy of access during the season of navigation, and supplied with an almost exhaustless amount of excellent pine timber. *Menekaunee* is at the extreme northern verge of the county, at the mouth of Menomonee River, and is already the seat of an important trade. The settlement (which also includes "Mission Point" and "Marinette") is scattered along the bank of the river for a distance of some two miles, and contains about 1,500 inhabitants.

"The extensive mills of the 'N. Y. Lumber Co.' are located at this point, and are well worth a trip from your city to see. Some idea may be formed of the vast amount of business done by this Company, when the fact is stated, that they had within their booms, at the commencement of the season, over *fifty acres of logs*. This vast amount of material will find its way into market, before the season closes, in the shape of good marketable lumber. This Company's Mills alone turn out over *half a million* of feet per week, all of which, I believe, goes to the Chicago market.

"But it must be borne in mind that this is but a fraction of the lumber which is made in this county. There are, besides the mills *here*, extensive lumbering establishments at Peshtigo, Oloton, Pensaukee, and Little Suamico, which probably turn out in the aggregate at least two million feet of lumber per week, or eighty millions per year—which added to the estimate for this place, makes a yearly product of one hundred and twenty millions. A fair amount of business, for a single county of not over 5,000 inhabitants."

MENOMONEE CITY, Oconto County, Wis., is a thriving settlement, situated on the west side of Green Bay, near the mouth of Menomonee River, containing 2,000 inhabitants, 5 large saw-mills, and several stores. The country to the west and north of this place is as yet a wilderness, inhabited only by a few roving Indians. The *Menomonee River* forms the boundary, in part, between Michigan and Wisconsin.

GREEN BAY TO LAKE SUPERIOR.

In regard to the route from Green Bay to Lake Superior, a distance of about 160 miles, the *Advocate* says:

"A road from Green Bay to the most southerly point of Keewenaw would be less than 200 miles in length, and while it would shorten the travel over the present route (by water) at least 100 miles, would open one of the most beautiful and fertile sections in the Union—a section which will remain unknown and unoccupied until such a road is opened by the government. The Lake Superior people need it most especially for procuring supplies, driving cattle, etc.

"The traveller finds the whole distance, to within a few miles of Lake Superior,

abounding in every resource which will make a country wealthy and prosperous. Clear, beautiful lakes are interspersed, and these have plenty of large trout and other fish. Water and water-powers are everywhere to be found, and the timber is of the best kind—maple groves, beech, oak, pine, etc. Nothing is now wanted but a few roads to open this rich country to the settler, and it will soon teem with villages, schools, mills, farming operations, and every industrial pursuit which the more southern portion of our State now exhibits."

PENSAUKEE, PESHTIGO, and other towns are springing up on the west shore of Green Bay, where are to be found numerous large lumber establishments situated on the streams running into the bay.

GENA, or MASON, situated on *Little Bay de Noc*, at the northern extremity of Green Bay, is a new and promising place. Steamers run to and from the town of Green Bay, connecting with mail stages running to Lake Superior. A *mail route* is now opened from GENA, situated at the head of Green Bay, to Marquette, L. S., a distance, by land, of about 50 miles. No doubt, ere long, a railroad will be constructed along the west shore of Green Bay, direct to Marquette, thus connecting Green Bay, Milwaukee, and Chicago with Lake Superior.

Route from Green Bay to Fond du Lac, Wisconsin.

There is now a railroad and steamboat route, extending from Green Bay to Appleton, Oshkosh, and Fond du Lac, situated at the head of Lake Winnebago, 60 miles distant, the latter passing through Fox River and the above beautiful sheet of water.

FOX or NEENAH RIVER rises in Marquette Co., Wis., and passing through Lake Winnebago, forms its outlet. This important stream is rendered navigable for steamers of a small class by means of dams and locks, forming, in connection with a short canal to the Wisconsin River, a direct water communication from Green Bay to the Mississippi River, a distance of about 200 miles. The rapids in the lower part of Fox River afford an immense water-power, while the upper section of country through which it flows, produces lumber and grain in great abundance. Here is a fall of 170 feet in the distance of 35 miles, before entering Lake Winnebago.

DE PERE, 5 miles above Green Bay, is a town of about 700 inhabitants, where is a fall of 8 feet, also a lock for the passage of steamers.

LITTLE KAUKAUNA, 11 miles, has a fall of 8 feet, with lock and dam.

WRIGHTSTOWN, 16 miles, is a small settlement, where is a steam saw-mill and other manufacturing establishments.

RAPID DE CROOPE, 2 miles further, is a steamboat landing. Here is a lock and dam, there being a fall of about 10 feet.

KAUKAUNA, 3 miles further, is a small village. Here are five locks, overcoming a fall of 60 feet.

LITTLE CHUTE, 25 miles from Green Bay, is a small French settlement, where is an old Roman Catholic Mission House. Here are four locks, there being a descent of 40 feet in the river.

APPLETON, Outaganie Co., Wis., is situated on Fox or Neenah River, 30 miles from its entrance into Green Bay, and five miles from Lake Winnebago, where are rapids called the *Grand Chute*. The river descends here about 30 feet in one mile and a half, affording an inexhaustible amount of water-power. Here are located three flouring-mills, six saw-mills, and several other extensive manufacturing establishments. This is the capital of the county, where is situated the *Lawrence University;* and it is no doubt destined to become a large manufacturing and commercial place, from the facilities which it possesses, by means of navigation and hy-

draulic power. Population, 4000. Steamers run south into Lake Winnebago, and north into Green Bay.

The approach to Appleton from Green Bay, by water, is most lovely and picturesque,—the river here winding through a rich section of country, clothed for several miles by a dense forest, extending to the very margin of the water. During the early autumn months the scene is truly gorgeous, the foliage presenting every variety of color.

MENASHA, 35 miles from Green Bay, is situated on an expansion of the river, here called *Lake Butte des Morts*, where is a lock and a canal of about one mile in length. Here are several large manufacturing establishments, and a population of about 2,500.

NEENAH, lying at the foot of Lake Winnebago, on the west shore, is a flourishing village, of about 2,500 inhabitants.

LAKE WINNEBAGO is a most beautiful sheet of water, being 32 miles long and about 12 miles wide, with bold land on the east shore, while on the west it seems elevated but a few feet above the waters of the lake. It abounds with several varieties of fish, of a fine flavor, affording rare sport to the angler. Steamers run through the Upper Fox or Wolf River, emptying into the lake at Oshkosh, for upwards of 100 miles, bringing down immense quantities of lumber and agricultural products.

The City of OSHKOSH, lying on the west side of Lake Winnebago, 20 miles north of Fond du Lac, is a large and flourishing place, being favorably situated at the mouth of Fox River on both sides of the stream. It now contains an active population of about 9,000 inhabitants. From its wharves steamers run to all the ports on the lake and Fox River, while the *Chicago and Northwestern Railway* extends northward to Green Bay. It contains the county buildings, 10 churches, several well-kept hotels, 100 stores of different kinds, besides steam grist-mills, steam saw-mills, iron foundries, cabinet-shops, and a great number of other manufacturing establishments. This is a great mart for lumber, being brought down the Fox or Wolf River for upwards of 100 miles, this stream flowing through a fine *pine region* of country, for which northern Wisconsin is justly celebrated.

FOND DU LAC, capital of Fond du Lac County, is a flourishing city favorably situated at the head of Lake Winnebago, 87 miles N. N. W. from Milwaukee, and 176 miles from Chicago, by the *Chicago and Northwestern Railway*, now finished through to Green Bay, a total distance of 242 miles. Here are located the county buildings, 8 churches, 4 banks, 6 public-houses, 100 stores of different kinds, a steam grist-mill, 10 steam saw-mills, a steam car factory, steam-engine manufactory, machine-shops, and various other manufacturing establishments. The lumber and produce business is very extensively carried on here, affording profitable returns. Fond du Lac is celebrated for its *fountains*, water being found of a pure quality by means of Artesian Wells, in which the city abounds.

The *Fox River Improvement* is a work of great magnitude, affording by means of locks and dams a water communication from Green Bay to Lake Winnebago, and thence south-westward through the Upper Fox river to Portage City, where, by means of a canal, it interlocks with the Wisconsin River, falling into the Mississippi at Prairie du Chien.

This enterprise is thus graphically described:

"'MEETING OF THE WATERS.'—A gentleman, recently from Green Bay, mentioned a curious fact a day or two since, illustrative of the results of the completion of the River Improvement. He saw lying at the docks in that place the steamer *Appleton Belle*, built at Pittsburgh, and the steamer *Gurdon Grant*, built at Philadelphia—points on opposite sides of the Alleghany Mountains, and on waters flowing on the one hand to the Atlantic, and on the other to the Mississippi and Gulf of Mexico. The *Belle* had sailed northward and westward through the Ohio, Mississippi, and Wisconsin; and the *Grant* in a contrary direction through the Delaware and Hudson, along the Erie Canal, and the chain of the Great Lakes. These are the victories of commerce, in which Wisconsin is playing a prominent part."

The TRIP FROM CHICAGO to MACKINAC, &c., connecting at the latter place with the *Green Bay* route, is fully described in another part of this work.

Ports of Lake Michigan situated on the East and South Shores.

Michigan City, Ind., situated at the extreme south end of Lake Michigan, is distant 45 miles from Chicago by water, and 228 miles from Detroit by railroad route. The *New Albany and Salem Railroad*, 228 miles in length, terminates at this place, connecting with the Michigan Central Railroad. Several plank roads also terminate here, affording facilities for crossing the extensive prairies lying in the rear. Here are several large storehouses situated at the mouth of Trail Creek, intended for the storage and shipment of wheat and other produce; 15 or 20 stores of different kinds, several hotels, and a branch of the State Bank of Indiana. It was first settled in 1831, with the expectation that it would become a great emporium of trade; but owing to the want of a good harbor, and the rapid increase of Chicago, the expectations of its founders have not been realized. It now contains about 3,000 inhabitants, and is steadily increasing in wealth and numbers.

NEW BUFFALO, Mich., lying 50 miles east of Chicago by steamboat route, is situated on the line of the Michigan Central Railroad, 218 miles west of Detroit. Here have been erected a light-house and pier, the latter affording a good landing for steamers and lake craft. The settlement contains two or three hundred inhabitants, and several stores and storehouses. It is surrounded by a light, sandy soil, which abounds all along the east and south shores of Lake Michigan.

ST. JOSEPH, Berrien Co., Mich., is advantageously situated on the east shore of Lake Michigan, at the mouth of St. Joseph River, 194 miles west of Detroit. Here is a good harbor, affording about 10 feet of water. The village contains about 1,000 inhabitants, and a number of stores and storehouses. An active trade in lumber, grain, and fruit is carried on at this place, mostly with the Chicago market, it being distant about 70 miles by water. Steamers of a small class run from St. Joseph to Niles and Constantine, a distance of 120 miles, to which place the St. Joseph River is navigable.

St. Joseph River rises in the southern portion of Michigan and Northern Indiana, and is about 250 miles long. Its general course is nearly westward; is very serpentine, with an equable current, and flowing through a fertile section of country, celebrated alike for the raising of grain and different kinds of fruit. There are to be found several flourishing villages on its banks. The principal are Constantine, Elkhart, South Bend, and Niles.

NILES, situated on St. Joseph River, is 26 miles above its mouth by land, and 191 miles from Detroit by railroad route. This is a flourishing village, containing about 3,000 inhabitants, five churches, three hotels, several large stores and flouring mills; the country around producing large quantities of wheat and other kinds of grain. A small class of steamers run to St. Joseph below and other places

above, on the river, affording great facilities to trade in this section of country.

SOUTH HAVEN, Van Buren Co., lies at the mouth of Black River.

NAPLES, Allegan Co., lies on the east side of Lake Michigan, near the mouth of the Kalamazoo River.

AMSTERDAM, Ottawa Co., is a small village lying near the Lake shore, about 20 miles south of Grand Haven.

HOLLAND, situated on *Black Lake*, a few miles above Amsterdam, is a thriving town, settled mostly by Hollanders. Here is a good and spacious harbor.

The counties of Berrien, Cass, Van Buren, Kalamazoo, Allegan, Kent, and Ottawa are all celebrated as a fruit-bearing region.

The Ports extending from Grand Haven to Saginaw Bay are fully described in another portion of this work, as well as the bays and rivers falling into Lakes Michigan and Huron.

————

Chicago, "THE GARDEN CITY," is advantageously situated on the west side of Lake Michigan, at the mouth of Chicago River, in N. lat. 41° 52', and W. long. from Greenwich 87° 35'. It is elevated eight to ten feet above the lake, which secures it from ordinary floods, and extends westward on both sides of the river, about two miles distant from its entrance into Lake Michigan, the front on the lake being three or four miles from north to south. The harbor has a depth of from twelve to fourteen feet of water, which makes it a commodious and safe haven; and it has been much improved artificially by the construction of piers, which extend on each side of the entrance of the river for some distance into the lake, to prevent the accumulation of sand upon the bar. The light-house is on the south side of the harbor, and shows a fixed light on a tower forty feet above the surface of the lake;

there is also a beacon light on the end of the pier. In a naval and military point of view, this is one of the most important ports on the upper lakes, and should be strongly defended, it being the "*Odessa*" of these inland seas.

The city contains an United States custom-house and post-office building, a court-house and jail, the county buildings, Rush Medical College, a commercial college, a marine hospital, market-houses, eighty churches, eight banks, several fire and marine insurance companies, and a number of large public houses; gas-works, and water-works. The manufacturing establishments of Chicago are numerous and extensive, consisting of iron-foundries and machine-shops, railroad car manufactory, steam saw, planing, and flouring mills, manufactories of agricultural implements, etc. Numerous steamers and propellers ply between this place and Saut Ste. Marie, Lake Superior ports, Collingwood, Goderich, Detroit, Buffalo, and the various intermediate ports. Population in 1860, 109,260, and is rapidly increasing in numbers and wealth.

The principal hotels are the *Adams House, Briggs House, Lake House, Sherman House,* and the *Tremont House,* besides many other public-houses of less note. Chicago also boasts of several benevolent and charitable institutions, educational and medical; also hospitals and dispensaries. Its commerce is already immense, and rapidly increasing.

EXTRACT FROM A LETTER DATED,

"CHICAGO, *June* 3, 1863.

" *To the Editors of the National Intelligencer.*

"This 'Garden City' of the Lakes yesterday and to-day, at the opening of the Ship-Canal Convention, presented a scene of which every loyal American might justly feel proud, be he a citizen of the

Atlantic States, of the great Valley of the Mississippi, or of the noble St. Lawrence valley, the waters of which here almost commingle, and no doubt are destined to be wedded, and to flow in unison from the cold waters of the Gulf of St. Lawrence to the warm waters of the Gulf of Mexico, carrying on their tide the rich products of the tropics, the cereals of the temperate zone, and the timber and other valuable commodities of the colder region of the American continent.

"Chicago, when fairly understood, in a commercial, climatic, and favored point of view, as regards water and land communication, has probably no equal on the face of the globe. Standing near the southern border of one of the five great Lakes of America, affording four or five thousand miles of inland ship navigation, and by means of the Erie Canal of New York, favored by an outport on the Atlantic ocean, it only now wants a *Ship Canal* to the Mississippi River to give it an outport on the Gulf of Mexico.

"As to Railroad facilities, no adequate idea can be conveyed.—the *iron bars* already extend to Montreal, Quebec, and Portland on the northeast; to Boston, New York, &c., on the east; to Charleston and Savannah on the southeast, and to Mobile and New Orleans on the south—being, ere long, no doubt destined to have railroad facilities with other cities to the Gold Regions and the Pacific ocean on the west; while northward a railroad line is finished to Green Bay, Wisconsin, and another line nearly completed to St. Paul, Minnesota.

"Look at a map of the United States and Canada, and see her favored position near the centre of the temperate zone; while east and west she lies centrally between the Atlantic and eastern slope of the Rocky Mountains. By nature she claims pre-eminence as a great inland city. Give her the boon she asks at this time, and the whole Republic will be benefited,

6

inasmuch as it will afford facilities to the far West, and the eastern portions of our wide-extended country. Like the Erie canal, it will even tend to lessen the price of provisions in foreign markets, and at the same time strengthen and enrich our own people, North and South, notwithstanding the latter are now in open rebellion."

"A late writer remarks, 'Chicago is most emphatically *the* city of the West; for when any town can justly claim to be the greatest market for grain, beef, pork, and lumber in the world, then we may confidently believe that all else that enters into the composition of a great city will gather there to build up, if not the greatest, one of the most important cities of the continent.'"

The *Illinois and Michigan Canal*, connecting Lake Michigan with Illinois River, which is sixty feet wide at the top, 6 feet deep, and 107 miles in length, including five miles of river navigation, terminates here, through which is brought a large amount of produce from the south and southwest; and the numerous railroads radiating from Chicago add to the vast accumulation which is here shipped for the Atlantic sea-board. Chicago being within a short distance of the most extensive coal-fields to be found in Illinois and the pineries of Michigan and Wisconsin, as well as surrounded by the finest grain region on the face of the globe, makes it the natural outlet for the varied and rich produce of an immense section of fertile country.

It is now proposed to construct a SHIP CANAL, connecting the waters *of the Mississippi River with Lake Michigan.*

RAILROADS DIVERGING FROM CHICAGO.

	Miles.
1. *Chicago, Alton, and St. Louis Railroad*	281
2. *Chicago, Burlington, and Quincy*	268
3. *Chicago and Rock Island*	182

	Miles.
4. *Chicago, Fulton, and Iowa* (finished)	249
5. *Chicago and North-Western* (Chicago to Green Bay)	242
6. *Chicago and Milwaukee*	85
7. *Cincinnati and Chicago Air Line*	280
8. *Galena and Chicago Union** (Chicago to Freeport)	121
9. *Illinois Central* (Chicago to Cairo)	365
10. *Michigan Central* (Detroit to Chicago)	284
11. *Michigan Southern and Northern Indiana* (Chicago to Toledo, Ohio)	243
12. *Pittsburgh, Fort Wayne, and Chicago*	468

* This road connects with the Illinois Central Railroad, running *westward* to Dunleith, opposite Du Buque, Iowa.
† *Illinois Central Railroad and its Branches:*
Cairo to Lasalle, Main Line........308 Miles.
Lasalle to Dunleith, Galena Branch....147 "
Chicago to Centralia, Chicago Branch..267 "

Total length ....722 Miles.

Distances from Chicago to Mackinac.

Running along the West Shore of Lake Michigan.

Ports.		Miles.
CHICAGO to Waukegan, Ill.		35
Kenosha, Wis.	16	51
Racine, Wis.	11	62
MILWAUKEE, Wis.	23	85
Fort Washington, Wis.	25	110
Sheboygan, Wis.	25	135
Manitowoc, Wis.	30	165
Two Rivers, Wis.	7	172
Kewaunee, Wis.	22	194
Anheepee, Wis.	11	205
Bayley's Harbor, Wis.	35	240
Death's Door.	20	260
(To Green Bay 80 miles.)		
Washington Harbor, Mich.	13	273
Rock Island, Mich.	7	280
Beaver Island, Mich.	67	347
Pt. Waugoshance, Mich.	30	377
Old Mackinac, Mich.	15	392
MACKINAC, Mich.	8	400

Mackinac to De Tour Passage	36
De Tour to Saut Ste. Marie	56
Mackinac to St. Clair River (Fort Gratiot)	240
Fort Gratiot to Detroit	60
Fort Gratiot to Saginaw City	150
Fort Gratiot to Goderich, C. W	60

NAVAL VESSELS ON THE LAKES.

MISSISSIPPI AND LAKE MICHIGAN CANAL.

A REPORT has recently been made in relation to the practicability, cost, and military and commercial advantages of opening a passage for gunboats and armed vessels from the Mississippi to the Lakes, by improving the navigation of the Illinois River, and enlarging the Illinois and Michigan Canal. The following is an extract from the above Report:

"The Great Lakes and the Mississippi River are among the grandest features of the geography of the globe. Their names are at once suggestive of commercial and agricultural wealth and national greatness. No such systems of internal navigation exist elsewhere in the world. The most careful and accurate statements of their present uses for commercial purposes are truly wonderful, while the magnificent future to which enlightened enterprise may lead, tasks the strongest imagination.

"The Mississippi system of navigable waters is variously estimated at from 10,000 to 20,000 miles. Its numerous ramifications penetrate a country of unrivalled fertility, and in many parts abounding in useful metals. On the Lakes, we have a coast of 3,500 miles. Their commerce is estimated at the value of $400,000,000, 'in articles of prime necessity to the inhabitants of the Eastern States, and to our foreign commerce.' That of the Mississippi, in peaceful times, is supposed to equal this. It is the union of these two mighty systems that we contemplate in the proposed improvement.

"For this purpose no other route exists comparable to the line now proposed, in the economy of cost of the improvement, or in general utility. It is one of nature's highways—one of the lines which she marks out for the guidance of the great emigrant movements of the race, and by which topography foretells the march of empire. The aboriginal savage travelled it by instinct, and now educated intelligence can find no better place for completing and uniting lines of travel and traffic embracing half a continent."

Estimate for a *Ship and Steamboat Canal* from Lake Michigan to the Illinois River, and the improvement of the Illinois River to the Mississippi River; the canal to be 160 feet wide on the bottom, sides protected with stone walls 10 feet high; the canal and river locks to be 350 feet long and 70 feet wide, with depth of water sufficient to pass steamboats and vessels drawing six feet of water; the canal to be supplied with water from Lake Michigan.

Chicago to Lockport, 29 miles.

The estimated cost of earth and rock excavation on the summit level from Chicago to Lockport, with walls on both sides 10 feet through the earth, is......................	$7,092,700
Lockport to Lasalle, 67 miles: The estimated cost of canal to Lake Joliet, and short canals at 16 locks, walled on both sides; also six stone dams, 600 feet long, eleven canal and five river locks, each 350 feet long and 70 feet wide—making 138 feet of lockage between Lockport and Lasalle—is	4,081,092
Lasalle to the Mississippi River, 220 miles: The cost of seven tree and crib dams, 900 feet long, the cribs to be filled with stone, and stone abutments; also seven stone locks, 350 feet long, and 70 feet wide, with entrances protected, and insuring a depth of water on all bars, to pass the largest class steamboats and vessels drawing six feet, will be......................	1,645,000
Add for bridges, right of way, engineering, contingencies, &c..............	578,032
Total......................	$18,846,824

U. S. NAVAL VESSELS ON LAKE ERIE, 1812 TO 1815.

Name.	Class.	Guns.	Where built, &c.	Date.
Amelia,	schooner	1	Erie, Pa.	1812
Ariel,	"	4	Purchased	1812
Caledonia,	brig	3	Captured	1812
Scorpion,	schooner	2	Purchased	1812
Somers,	"	2	"	1812
Trippe,	sloop	1	"	1812
Tigress,	schooner	1	Erie, Pa.	1812
Lawrence,	brig	20	"	1813
Niagara,	"	20	"	1813
Ohio,	schooner	1	"	1813
Porcupine,	"	1	"	1813
Ghent,	"	1	"	1815

Total Guns, 57

NOTE.—At the conclusion of the war with Great Britain, this squadron, with the exception of two small vessels, and the prizes captured by the above fleet, under Com. Perry, Sept. 10, 1813, were dismantled and laid up at Erie, Penn., and all subsequently condemned and sold, some having been sunk, with a view to their better preservation.

In 1844, the U. S. steamer MICHIGAN, 583 tons, was built at Erie, Pa., and is now in commission on the Upper Lakes.

ROUTE FROM CHICAGO TO MACKINAC AND SAUT STE. MARIE.

On starting from the steamboat wharf near the mouth of the Chicago River, the Marine Hospital and depot of the Illinois Central Railroad are passed on the right, while the Lake House and lumber-yards are seen on the left or north side of the stream. The government piers, long wooden structures, afford a good entrance to the harbor; a light-house has been constructed on the outer end of the north pier, to guide vessels to the port.

The basin completed by the Illinois Central Railroad to facilitate commerce is a substantial work, extending southward for nearly half a mile. It affords ample accommodation for loading and unloading vessels, and transferring the freight to and from the railroad cars.

The number of steamers, propellers, and sailing vessels annually arriving and departing from the harbor of Chicago is very great; the carrying trade being destined to increase in proportionate ratio with the population and wealth pouring into this favored section of the Union.

On reaching the green waters of Lake Michigan, the city of Chicago is seen stretching along the shore for four or five miles, presenting a fine appearance from the deck of the steamer. The entrance to the harbor at the bar is about 200 feet wide. The bar has from ten to twelve feet water, the lake being subject to about two feet rise and fall. The steamers bound for Milwaukee and the northern ports usually run along the west shore of the lake within sight of land, the banks rising from thirty to fifty feet above the water.

LAKE MICHIGAN is about seventy miles average width, and 340 miles in extent from Michigan City, Ind., on the south, to the Strait of Mackinac on the north; it presents a great expanse of water, now traversed by steamers and other vessels of a large class, running to the Saut Ste. Marie and Lake Superior; to Collingwood and Goderich, Can.; to Detroit, Mich.; to Cleveland, Ohio, and to Buffalo, N. Y. From Chicago to Buffalo the distance is about 1,000 miles by water; while from

Chicago to Superior City, at the head of Lake Superior, or Fond du Lac, the distance is about the same, thus affording two excursions of 1,000 miles each, over three of the great lakes or inland seas of America, in steamers of from 1,000 to 2,000 tons burden. During the summer and early autumn months the waters of this lake are comparatively calm, affording safe navigation. But late in the year, and during the winter and early spring months, the navigation of this and the other great lakes is very dangerous.

WAUKEGAN, Lake Co., Ill. 36 miles north of Chicago, is handsomely situated on elevated ground, gradually rising to 50 or 60 feet above the water. Here are two piers, a light-house, several large storehouses. and a neat and thriving town containing about 5,000 inhabitants, six churches, a bank, several well-kept hotels, thirty stores, and two steam-flouring mills.

KENOSHA, Wis., 52 miles from Chicago, is elevated 30 or 40 feet above the lake. Here are a small harbor, a light-house, storehouses, mills. etc. The town has a population of about 6,000 inhabitants, surrounded by a fine back country. Here is a good hotel, a bank, several churches, and a number of stores and manufacturing establishments doing a large amount of business. The *Kenosha and Rockford Railroad*, 73 miles, connects at the latter place with a railroad running to Madison, the capital of the State, and also to the Mississippi River.

The City of RACINE, Wis., 62 miles from Chicago and 23 miles south of Milwaukee, is built on an elevation some forty or fifty feet above the surface of the lake. It is a beautiful and flourishing place. Here are a light-house, piers, storehouses, etc., situated near the water, while the city contains some fine public buildings and private residences. The population is about 10,000, and is rapidly increasing. Racine is the second city in the State in

commerce and population, and possesses a fine harbor. Here are located the county buildings, fourteen churches, several hotels, *Congress Hall* being the largest; elevators, warehouses, and numerous stores of different kinds.

The *Racine and Mississippi Railroad* extends from this place to the Mississippi River at Savanna, 142 miles. The Chicago and Milwaukee Railroad also runs through the town, near the Lake Shore.

MILWAUKEE HARBOR.

Milwaukee, "THE CREAM CITY," 86 miles from Chicago, by railroad and steamboat route, is handsomely situated on rising ground on both sides of the Milwaukee River, at its entrance into Lake Michigan. In front of the city is a bay or indentation of the lake, affording a good harbor, except in strong easterly gales. The harbor is now being improved, and will doubtless be rendered secure at all times of the season. The river affords an extensive water-power, capable of giving motion to machinery of almost any required amount. The city is built upon beautiful slopes, descending toward the river and lake. It has a United States Custom House and Post-Office building; a court house, city hall, a United States land-office, the University Institute, a college for females, three academies, three orphan asylums, forty-five churches, several well-kept hotels, the *Newhall House* and the *Walker House* being the most frequented; seven banks, six insurance companies, a Chamber of Commerce, elevators, extensive ranges of stores, and several large manufacturing establishments. The city is lighted with gas, and well supplied with good water. Its exports of lumber, agricultural produce, etc. are immense, giving profitable employment to a large number of steamers and other lake craft, running to different ports on the Upper Lakes, Detroit, Buffalo, etc. The growth of this city has been astonishing; twenty years since its site was a wilderness; now it contains over 50,000 inhabitants, and of a class inferior to no section of the Union for intelligence, sobriety, and industry.

The future of Milwaukee it is hard to predict; here are centring numerous railroads finished and in course of construction, extending south to Chicago, west to the Mississippi River, and north to Lake Superior, which, in connection with the Detroit and Milwaukee Railroad, terminating at Grand Haven, 85 miles distant by water, and the lines of steamers running to this port, will altogether give an impetus to this favored city, blessed with a good climate and soil, which the future alone can reveal.

During the past few years an unusual number of fine buildings have been erected, and the commerce of the port has amounted to $60,000,000. The bay of Milwaukee offers the best advantages for the construction of a harbor of refuge of any point on Lake Michigan. The city has expended over $100,000 in the construction of a harbor; this needs extension and completion, which will no doubt be effected.

The approach to Milwaukee harbor by water is very imposing, lying between two headlands covered with rich foliage, and dotted with residences indicating comfort and refinement not to be exceeded on the banks of the Hudson or any other body of water in the land. This city, no doubt, is destined to become the favored residence of opulent families, who are fond of congregating in favored localities.

THE GRANARIES OF MINNESOTA AND WISCONSIN.—The La Crosse *Democrat* speaks as follows of the great strides of agriculture in a region which ten years ago was a wilderness. It says:

"We begin to think that the granaries of Minnesota and Northwestern Wisconsin will never give out; there is no end to the amount, judging from the heavy loads the steamers continually land at the depot of the La Crosse and Milwaukee Railroad. Where does it all come from? is the frequent inquiry of people. We can hardly tell. It seems impossible that there can be much more left, yet steamboat men tell us that the grain is not near all hauled to the shipping points on the river. What will this country be ten years hence, at this rate? Imagine the amount of transportation that will become necessary to carry the produce of the upper country to market. It is hard to state what will be the amount of shipments of grain this season (1863), but it will be well into the millions."

RAILROADS RUNNING FROM MILWAUKEE.

Detroit and Milwaukee (Grand Haven to Detroit, 189 miles), connecting with steamers on Lake Michigan.

La Crosse and Milwaukee, 200 miles, connecting with steamers on the Upper Mississippi.

Milwaukee and Prairie du Chien, 192 miles, connecting with steamers on the Mississippi River.

Milwaukee and Horicon, 93 miles.

Milwaukee and Western, 71 miles.

Milwaukee and Chicago, 85 miles; also, the River and Lake Shore City Railway, running from the entrance of the harbor to different parts of the city.

PORT WASHINGTON, Ozaukee Co., Wis., 25 miles north of Milwaukee, is a flourishing place, and capital of the county. The village contains, besides the public buildings, several churches and hotels, twelve stores, three mills, an iron foundry, two breweries, and other manufactories. The population is about 2,500. Here is a good steamboat landing, from which large quantities of produce are annually shipped to Chicago and other lake ports.

SHEBOYGAN, Wis., 50 miles north of Milwaukee and 130 miles from Chicago, is a thriving place, containing about 6,000 inhabitants. Here are seven churches, several public-houses and stores, together with a light-house and piers ; the harbor being improved by government works. Large quantities of lumber and agricultural products are shipped from this port. The country in the interior is fast settling with agriculturists, the soil and climate being good. A railroad nearly completed runs from this place to FOND DU LAC, 42 miles west, lying at the head of Lake Winnebago.

MANITOUWOC, Wis., 70 miles north of Milwaukee and 33 miles east from Green Bay, is an important shipping port. It contains about 2,500 inhabitants; five churches, several public-houses, twelve stores, besides several storehouses; three steam saw-mills, two ship-yards, lighthouse, and pier. Large quantities of lumber are annually shipped from this port. The harbor is being improved so as to afford a refuge for vessels during stormy weather.

"Manitouwoc is the most northern of the harbors of Lake Michigan improved by the United States government. It derives additional importance from the fact that, when completed, it will afford the first point of refuge from storms for shipping bound from any of the other great lakes to this, or to the most southern ports of Lake Michigan."

TWO RIVERS, Wis., seven miles north from Manitouwoc, is a new and thriving place at the entrance of the conjoined streams (from which the place takes its name) into Lake Michigan. Two piers are here erected, one on each side of the river; also a ship-yard, an extensive leather manufacturing company, chair and pail factory, and three steam saw-mills. The village contains about 2,000 inhabitants.

KEWAUNEE, Wis., 25 miles north of Two Rivers and 102 miles from Milwaukee, is a small shipping town, where are situated several saw-mills and lumber establishments. Green Bay is situated about 25 miles due west from this place.

AHNEEPEE, 12 miles north of Kewaunee, is a lumbering village, situated at the mouth of Ahneepee, containing about 1,000 inhabitants. The back country here assumes a wild appearance, the forest trees being mostly pine and hemlock.

GIBRALTAR, or BAILEY'S HARBOR, is a good natural port of refuge for sailing craft when overtaken by storms. Here is a settlement of some 400 or 500 inhabitants, mostly being engaged in fishing and lumbering.

PORT DES MORTS or DEATH'S DOOR, the entrance to Green Bay, is passed 20 miles north of Bailey's Harbor. *Detroit Island* lying to the northward.

POTTOWATOMEE, or WASHINGTON ISLAND, is a fine body of land attached to the State of Michigan ; also, Rock Island, situated a short distance to the north. (*See route to Green Bay, &c.*).

On leaving *Two Rivers*, the steamers passing through the Straits usually run for the Manitou Islands, Mich., a distance of about 100 miles. Soon after the last vestige of land sinks below the horizon on the west shore, the vision catches the dim outline of coast on the east or Michigan shore at *Point aux Bec Scies*, which is about 30 miles south of the Big Manitou Island. From this point, passing northward by *Sleeping Bear Point*, a singular shaped headland looms up to the view. It is said to resemble a sleeping bear. The east shore of Lake Michigan presents a succession of high sand-banks for many miles, while inland are numerous small bays and lakes.

LITTLE, or SOUTH MANITOU ISLAND, 260 miles from Chicago, and 110 miles from Mackinac, lies on the Michigan side of the lake, and is the first island encountered on proceeding northward from Chicago. It rises abruptly on the west shore 2 or 300 feet from the water's edge, sloping toward the east shore, on which is a light-house and a fine harbor. Here steamers stop for wood. BIG or NORTH MANITOU is nearly twice as large as the former island, and contains about 14,000 acres of land. Both islands are settled by a few families, whose principal occupation is fishing and cutting wood for the use of steamers and sailing vessels.

FOX ISLANDS, 50 miles north from South Manitou, consist of three small islands lying near the middle of Lake Michigan, which is here about 60 miles wide. On the west is the entrance to Green Bay, on the east is the entrance to Grand Traverse Bay, and immediately to the north is the entrance to Little Traverse Bay.

GREAT and LITTLE BEAVER Islands lying about midway between the Manitou Islands and Mackinac, are large and fertile bodies of land, formerly occupied by Mormons, who had here their most eastern settlement.

GARDEN and HOG Islands are next pass-ed before reaching the Strait of Mackinac, which, opposite Old Fort Mackinac, is about six miles in width. The site of Old Fort Mackinac is on the south main or Michigan shore, directly opposite Point Ste. Ignace, on the north main shore. *St. Helena Island* lies at the entrance of the strait from the south, distant about fifteen miles from Mackinac.

OLD FORT MACKINAC,* now called *Mackinac City*, is an important and interesting location; it was formerly fortified and garrisoned for the protection of the strait and this section of country, when inhabited almost exclusively by various tribes of Indians. This place can be easily reached by sail-boat from the island of Mackinac.

PTE. LE GROS CAP, lying to the west of old Fort Mackinac, is a picturesque headland well worthy of a visit.

The STRAIT OF MACKINAC is from five to twenty miles in width, and extends east and west about forty miles, embosoming several important islands besides Mackinac Island, the largest being BOIS BLANC ISLAND, lying near the head of Lake Huron. Between this island and the main north shore the steamer GARDEN CITY was wrecked, May 16, 1854; her upper works were still visible from the deck of the passing steamer in the fall of the same year.

GROSSE ILE ST. MARTIN and Ile St. Martin lie within the waters of the strait, eight or ten miles north of the island of Mackinac. In the neighborhood of these different islands are the favorite fishing-grounds both of the Indian and the "pale face."

Mackinac, the Town and Fortress, is most beautifully situated on the east shore of the island, and extends for a distance of about one mile along the water's edge, and has a fine harbor protected by a

* Settled by the French under Father Marquette in 1670.

water battery.   This important island and fortress is situated in N. lat. 45° 54′, W. lon. 84° 30′ from Greenwich, being seven degrees thirty minutes west from Washington.   It is 350 miles north from Chicago, 100 miles south of Saut Ste. Marie by the steamboat route, and about 300 miles northwest from Detroit.   *Fort Mackinac*, garrisoned by U. States troops, stands on elevated ground, about 200 feet above the water, overlooking the picturesque town and harbor below.   In the rear, about half a mile distant stand the ruins of old *Fort Holmes*, situated on the highest point of land, at an elevation of 320 feet above the water, affording an extensive view.

The town contains two churches, five hotels, ten or twelve stores, 100 dwelling-houses, and about 700 inhabitants.   The climate is remarkably healthy and delightful during the summer months, when this favored retreat is usually thronged with visitors from different parts of the Union, while the Indian warriors, their squaws and their children, are seen lingering around this their favorite island and fishing-ground.

The Island of MACKINAC, lying in the Strait of Mackinac, is about three miles long and two miles wide.   It contains many deeply interesting points of attraction in addition to the village and fortress; the principal natural curiosities are known as the Arched Rock, Sugar Loaf, Lover's Leap, Devil's Kitchen, Robinson's Folly, and other objects of interest well worthy the attention of the tourist.   The *Mission House* and *Island House* are the principal hotels, while there are several other good public-houses for the accommodation of visitors.

ISLAND OF MACKINAC.—The view given represents the Island, approaching from the eastward.   "A cliff of limestone, white and weather-beaten, with a narrow alluvial plain skirting its base, is the first thing which commands attention ;" but, on nearing the harbor, the village (2), with its many picturesque dwellings, and the fortress (3), perched near the summit of the Island, are gazed at with wonder and delight.   The promontory on the left is called the "Lover's Leap" (1), skirted by a pebbly beach, extending to the village.   On the right is seen a bold rocky precipice, called "*Robinson's Folly*" (5), while in the same direction is a singular peak of nature called the "*Sugar Loaf.*" Still farther onward, the "*Arched Rock,*" and other interesting sights, meet the eye of the explorer, affording pleasure and delight, particularly to the scientific traveller and lover of nature.   On the highest ground, elevated 320 feet above the waters of the Strait, is the signal station (4), situated near the ruins of old *Fort Holmes*.

The settlement of this Island was commenced in 1764.   In 1793 it was surrendered to the American government ; taken by the British in 1812 ; but restored by the treaty of Ghent, signed in Nov., 1814

The Lover's Leap.—MACKINAC ISLAND.—The huge rock called the "Lover's Leap," is situated about one mile west of the village of Mackinac. It is a high perpendicular bluff, 150 to 200 feet in height, rising boldly from the shore of the Lake. A solitary pine-tree formerly stood upon its brow, which some Vandal has cut down.

Long before the pale faces profaned this island home of the Genii, Me-che-ne-mock-e-nung-o-qua, a young Ojibway girl, just maturing into womanhood, often wandered there, and gazed from its dizzy heights and witnessed the receding canoes of the large war parties of the combined bands of the Ojibways and Ottawas, speeding South, seeking for fame and scalps.

It was there she often sat, mused, and hummed the songs Ge-niw-e-gwon loved; this spot was endeared to her, for it was there that she and Ge-niw-e-gwon first met and exchanged words of love, and found an affinity of souls or spirits existing between them. It was there she often sat and sang the Ojibway love song—

"Mong-e-do-gwain, in-de-nain-dum,
Mong-e-do-gwain, in-de-nain-dum;
Wain-shung-ish-ween, neen-e-mo-shane,
Wain-shung-ish-ween, neen-e-mo-shane,
A-nee-wau-wau-sau-bo-a-zode,
A-nee-wau-wau-sau-bo-a-zode."

I give but one verse, which may be translated as follows:

A loon, I thought was looming,
A loon, I thought was looming;
Why! it is he, my lover,
Why! it is he, my lover.
His paddle in the waters gleaming,
His paddle in the waters gleaming.

From this bluff she often watched and listened for the return of the war parties, for amongst them she knew was Ge-niw-e-gwon; his head decorated with war-eagle plumes, which none but a brave could sport. The west wind often wafted far in advance the shouts of victory and death, as they shouted and sang upon leaving Pe-quot-e-uong (old Mackinac), to make the traverse to the Spirit, or Fairie Island.

One season, when the war party returned, she could not distinguish his familiar and loved war-shout. Her thinking spirit, or soul (presentiment) told her that he had gone to the Spirit Land of the west. It was so, an enemy's arrow had pierced his breast, and after his body was placed leaning against a tree, his face fronting his enemies he died; but ere he died he wished the mourning warriors to remember him to the sweet maid of his heart. Thus he died far away from home and the friends he loved.

Me-che-ne-mock-e-nung-o-qua's heart hushed its beatings, and all the warm emotions of that heart were chilled and dead. The moving, living spirit or soul of her beloved Ge-niw-e-gwon she witnessed, continually beckoning her to follow him to the happy hunting grounds of spirits in the west—he appeared to her in human shape, but was invisible to others of his tribe.

One morning her body was found mangled at the foot of the bluff. The soul had thrown aside its covering of earth, and had gone to join the spirit of her beloved Ge-niw-e-gwon, to travel together to the land of spirits, realizing the glories and bliss of a future, eternal existence.

Yours, &c.,
WM. M. J * * * * * *

ALTITUDE OF VARIOUS POINTS ON ISLAND OF MACKINAC.

Localities.	Above Lake Huron.	Above the Sea.
Lake Huron	000 feet.	574 feet.
Fort Mackinac	150 "	724 "
Old Fort Holmes	315 "	889 "
Robinson's Folly	128 "	702 "
Chimney Rock	181 "	705 "
Top of Arched Rock	140 "	714 "
Lover's Leap	145 "	719 "
Summit of Sugar Loaf	284 "	859 "
Principal Plateau of Mackinac Island	160 "	734 "
Upper Plateau	300 "	874 "
La Cloche Mountain, north side Lake Huron, C. W.	1,200 "	1,774 "

ARCHED ROCK.—Mackinac.

The whole Island of Mackinac is deeply interesting to the scientific explorer, as well as to the seeker of health and pleasure. The following extract, illustrated by an engraving. is copied from "FOSTER and WHITNEY'S *Geological Report*" of that region:

"As particular examples of denuding action on the island, we would mention the 'Arched Rock' and the 'Sugar Loaf.' The former, situated on the eastern shore, is a feature of great interest. The cliffs here attain a height of nearly one hundred feet, while at the base are strewn numerous fragments which have fallen from above. The *Arched Rock* has been excavated in a projecting angle of the limestone cliff, and the top of the span is about ninety feet above the lake-level, surmounted by about ten feet of rock. At the base of a projecting angle, which rises up like a buttress, there is a small opening, through which an explorer may pass to the main arch, where, after clambering over the steep slope of debris and the projecting edges of the strata, he reaches the brow of the cliff.

"The beds forming the summit of the arch are cut off from direct connection with the main rock by a narrow gorge of no great depth. The portion supporting the arch on the north side, and the curve of the arch itself, are comparatively fragile, and cannot, for a long period, resist the action of rains and frosts, which, in this latitude, and on a rock thus constituted, produce great ravages every season. The arch, which on one side now connects this abutment with the main cliff, will soon be destroyed, as well as the abutment itself, and the whole be precipitated into the lake.

"It is evident that the denuding action reducing such an opening, with other attendant phenomena, could only have operated while near the level of a large body of water like the great lake itself; and we find a striking similarity between the denuding action of the water here in time past, and the same action as now manifested in the range of the *Pictured Rocks* on the shores of Lake Superior. As an interesting point in the scenery of this island, the Arched Rock attracts much attention, and in every respect is worthy of examination." (*See Engraving*.)

Other pictures[que] objects of great interest, besides those enumerated above, occur at every turn on roving about this enchanting island, where the pure, bracing air and clear waters afford a pleasurable sensation, difficult to be described unless visited and enjoyed.

The bathing in the pure waters of the Strait at this place is truly delightful, affording health and vigor to the human frame.

The Island of Mackinac.

ROMANTIC AND PICTURESQUE APPEARANCE OF THE ISLAND AND SURROUNDING COUNTRY—ITS PURITY OF ATMOSPHERE—A MOONLIGHT EXCURSION, &c., &c.

——"From whose rocky turrets battled high,
Prospect immense spread out on all sides round;
Lost now between the welkin and the main,
Now walled with hills that slept above the storm,
Most fits such a place for musing men;
Happiest, sometimes, when musing without aim."
[POLLOK.

In this Northern region, Nature has at last fully resumed her green dress. Flowers wild, but still beautiful, bloom and disappear in succession. Birds of various hues have returned to our groves, and welcome us as we trace these shady walks. "In all my wand'rings round this world of care," I have found no place wherein the climate, throughout the summer season, seems to exercise on the human constitution a more beneficial influence than on this Island. In other parts of this country and in Europe, the places of *Resort* are beautiful, indeed; but a certain oppressiveness there at times pervades the air, that a person even with the best health in the world, feels a lassitude creeping through his frame. Here, we seldom, if ever, experience such a feeling from this cause. For the western breeze even in the hottest days passing over this island, keeps the air cool, and, especially if proper exercise be taken by walking or riding, one feels a bracing up, a certain buoyancy of spirits that is truly astonishing.

Ye inhabitants of warm latitudes, who pant in cities for a breath of cool air, fly to this isle for comfort. Ye invalid, this is the place in which to renovate your shattered constitution. The lovers of beautiful scenery or the curious in nature, and the artist, whose magic pencil delights to trace nature's lineaments, need not sigh for the sunny clime of Italy for subjects on which to feed the taste and imagination.

This island is intersected by fine carriage roads, shaded here and there by a young growth of beech, maple, and other trees. On the highest part of it, about 300 feet, are the ruins of Old Fort Holmes. From this point of elevation, the scenery around is extensive and beautiful. In sight, are some localities connected with "the tales of the times of old," both of the savage and the civilized. Looking westwardly, and at the distance of about four miles across an arm of Lake Huron, is Point St. Ignace, which is the southernmost point of land, of the greater portion of the Upper Peninsula. Immediately south of it are the "Straits of Mackinac," which separating the Northern and Southern Peninsulas from each other, are about four miles wide. On the south shore, may still be seen traces of Old Fort Mackinac, which is well known in history as having been destroyed by Indians, in 1763, at the instigation of Pontiac, an Indian Chief. Turning our gaze southeastwardly, we see the picturesque "Round Island," as it were at our feet. And further on, is "Bois-Blanc Island," stretching away with its winding shores, far into Lake Huron. Look to the east, and there stands this inland sea, apparently "boundless and deep," and "pure as th' expanse of heaven." Directly north from our place of observation, are the "Islands of St. Martin;" while beyond them in the Bay, are two large rivers—the Pine, and Carp Rivers. And lastly, casting our eyes towards the northwest, we see on the main land the two "Sitting Rabbits;" being two singular looking hills or rocks, and so called by the Indians from some resemblance at a distance to rabbits in a sitting posture. As a whole, this scenery presents, hills, points of land jutting into the lake, and "straits," bays, and islands. Here, the lake contracts itself into narrow channels, or straits, which at times are whitened by numerous sails of commerce; and there, it spreads itself away as far as the eye can reach. And, while contemplating this scene, perhaps a dark column of smoke, like the Genii in the Arabian Tales, may be seen rising slowly out of the bosom of Lake Huron, announcing the approach of the Genii of modern days, the Steamboat! Let us descend to the shore.

It is evening! The sun, with all his glory has disappeared in the west; but the moon sits in turn the arbitress of heaven. And now—

"How sweet the moonlight sleeps upon this bank;
Here will we sit, and let the sounds of music
Creep in our ears: soft stillness and the night,
Becomes the touches of sweet harmony."

Such a moonlight night I once enjoyed. The hum of day-life had gradually subsided, and there was naught to disturb the stillness of the hour, save the occasional laughter of those who lingered out in the open air. In the direction of the moon, and on the Lake before me, there was a broad road of light trembling upon its bosom. A few moments more, two small boats with sails up to catch the gentle breeze, were seen passing and re-passing this broad road of light. Then the vocal song was raised on the waters, and woman's voice was borne on moonlight beam to the listening ear in the remotest shades. The voices became clearer and stronger as the boats approached nearer; then, again, dying away in the distance, seemed to be merged with the mellow rays of the moon. But let us leave poetry and fancy aside, and come to matters of fact, matters of accommodation, prepared for those who may favor our island with their visits this summer.

There are several large hotels, with attentive hosts, ever ready to contribute towards the comforts of their visitors. Walking, riding, fishing, shooting, and sailing can be here pursued with great benefit to health. We have billiard-rooms and bowling-alleys; in the stores are found Indian curiosities; and, perhaps, the Indians themselves, who resort to this island on business, may be curiosities to those who have never seen them; they are the true "native Americans," the *citizens* of this North American Republic.

ROUND ISLAND is a small body of land lying a short distance southeast of Mackinac, while BOIS BLANC ISLAND is a large body of land lying still farther in the distance, in the Straits of Mackinac.

ST. MARTIN'S BAY, and the waters contiguous, lying north of Mackinac, afford fine fishing grounds, and are much resorted to by visitors fond of aquatic sports. *Great St. Martin's* and *Little St. Martin's Islands* are passed before entering the bay, and present a beautiful appearance.

CARP and PINE rivers are two small streams entering into St. Martin's Bay, affording an abundance of brook trout of a large size. From the head of the above bay to the foot of Lake Superior, is only about 30 miles in a northerly direction, passing through a wilderness section of country, sparsely inhabited by Indians, who have long made this region their favored hunting and fishing grounds.

POINT DE TOUR, 36 miles east from Mackinac, is the site of a light-house and settlement, at the entrance of St. Mary's River, which is here about half a mile in width; this passage is also called the West Channel. At a distance of about two miles above the Point is a new settlement, where have been erected a steamboat pier, a hotel, and several dwellings.

DRUMMOND ISLAND, a large and important body of land belonging to the United States, is passed on the right, where are to be seen the ruins of an old fort erected by the British. On the left is the mainland of Northern Michigan. Ascending St. Mary's River, next is passed ROUND or PIPE ISLAND, and other smaller islands on the right, presenting a beautiful appearance, most of them belonging to the United States.

ST. JOSEPH ISLAND, 10 miles above Point de Tour, is a large and fertile island belonging to Canada. It is about 20 miles long from east to west, and about 15 miles broad, covered in part with a heavy growth of forest-trees. Here are seen the ruins of an old fort erected by the British, on a point of land commanding the channel of the river.

CARLTONVILLE is a small settlement on the Michigan side of the river, 12 miles above the De Tour. Here is a steam saw-mill and a few dwelling-houses.

LIME ISLAND is a small body of land belonging to the United States, lying in the main channel of the river, about 12 miles from its mouth. The channel here forms the boundary between the United States and Canada.

MUD LAKE, as it is called, owing to its waters being easily riled, is an expansion of the river, about five miles wide and ten miles long, but not accurately delineated on any of the modern maps, which appear to be very deficient in regard to St. Mary's River and its many islands—presenting at several points most beautiful river scenery. In the St. Mary's River there are about fifty islands belonging to the United States, besides several attached to Canada.

NEBISH ISLAND, and *Sailor's Encampment*, situated about half way from the Point to the Saut, are passed on the left while sailing through the main channel.

SUGAR ISLAND, a large body of fertile land belonging to the United States, is reached about 30 miles above Point de Tour, situated near the head of St. Joseph Island. On the right is passed the *British* or *North Channel*, connecting on the east with Georgian Bay. Here are seen two small rocky islands belonging to the British Government, which command both channels of the river.

The *Nebish Rapids* are next passed by the ascending vessel, the stream here running about five knots per hour. The mainland of Canada is reached immediately above the rapids, being clothed with a dense growth of forest-trees of small size. To the north is a dreary wilderness, extending through to Hudson Bay, as yet almost wholly unexplored and unknown, except to the Indian or Canadian hunter.

LAKE GEORGE, twenty miles below the Saut, is another expansion of the river, being about five miles wide and eight miles long. Here the channel is only from eight to ten feet in depth for about one mile, forming a great impediment to navigation.*

CHURCH'S LANDING, on Sugar Island, twelve miles below the Saut, is a steamboat landing; opposite it is SQUIRREL ISLAND, belonging to the Canadians. This is a convenient landing, where are situated a store and dwelling. The industrious occupants are noted for the making of *raspberry jam*, which is sold in large quantities, and shipped to Eastern and Southern markets.

Garden River Settlement is an Indian

* A new channel has been formed, by dredging, which gives a greater depth of water.

village ten miles below the Saut, on the Canadian shore. Here are a missionary church and several dwellings. surrounded by grounds poorly cultivated. fishing and hunting being tue main employment of the Chippewa Indians who inhabit this section of country. Both sides of the river abound in wild berries of good flavor, which are gathered in large quantities by the Indians, during the summer months.

Extract from a letter dated SAUT STE. MARIE, Sept., 1854:

"The scenery of the St. Mary's River seems to grow more attractive every year. There is a delicious freshness in the countless evergreen islands that dot the river in every direction, from the Falls to Lake Huron, and I can imagine of no more tempting retreats from the dusty streets of towns, in summer, than these islands; I believe the time will soon come when neat summer cottages will be scattered along the steamboat route on these charming islands. A summer could be delightfully spent in exploring for new scenery and in fishing and sailing in these waters.

"And Mackinac, what an attractive little piece of *terra firma* is that island—half ancient, half modern! The view from the fort is one of the finest in the world. Perched on the brink of a precipice some two hundred feet above the bay—one takes in at a glance from its walls the harbor, with its numerous boats and the pretty village; and the whole rests on one's vision more like a picture than a reality. Every thing on the island is a curiosity; the roads or streets that wind around the harbor or among the grove-like forests of the island are naturally pebbled and macadamized; the buildings are of every style, from an Indian lodge to a fine English house. The island is covered with charming natural scenery, from the pretty to the grand, and one may spend weeks constantly finding new objects of interest and new scenes of beauty. It is unnecessary to particularize—every visitor will find them, and enjoy the sight more than any description.

"The steamers all call there, on their way to and from Chicago, and hundreds of small sail vessels, in the fishing trade, have here their head-quarters. Drawn upon the pebbled beach or gliding about the little bay are bark canoes and the far-famed 'Mackinac boats,' without number. These last are the perfection of light sail-boats, and I have often been astonished at seeing them far out in the lake, beating up against winds that were next to gales. Yesterday the harbor was thronged with sail boats and vessels of

every description, among the rest were the only two iron steamers that the United States have upon all the lakes, the 'Michigan' and the 'Surveyor,' formerly called the 'Abert,' employed in the Coast Survey.

"For a wonder, Lake Huron was calm and at rest for its entire length, and the steamer 'Northerner' made a beautiful and quick passage from Mackinac to this place. The weather continues warm and dry, and hundreds are regretting they have so early left the Saut and Mackinac, and we believe you will see crowds of visitors yet.   JAY."

St. Mary's River.

By a careful examination of the Government Charts of the Straits of Mackinac and River Ste. Marie, published in 1857, it appears that the *Point De Tour Light-House* is situated in 45° 57' N. Lat., being 36 miles to the eastward of Fort Mackinac. The width of the De Tour passage is about one mile, with a depth of water of 100 feet and upwards, although but 50 feet is found off the light, as you run into Lake Huron. *Drummond Island,* attached to the United States, lies on the east, while the main shore of Michigan lies to the west of the entrance. *Pipe Island,* 4 miles, is first passed on ascending the stream, and then *Lime Island,* 6 miles further. *St. Joseph's Island,* with its old fort, attached to Canada, lies 8 miles from

the entrance. *Potagannissing Bay,* dotted with numerous small islands, mostly belonging to the United States, is seen lying to the eastward, communicating with the North Channel. *Mud Lake,* 6 miles further, is next entered, having an expanse of about 4 miles in width, when *Sailor's Encampment Island* is reached, being 20 miles from Lake Huron. The head of St. Joseph's and part of *Sugar Island* are reached 26 miles northward from the De Tour, where diverges the Canadian or North Channel, running into the Georgian Bay; this channel is followed by the Canadian steamers. The *Nebish Rapids* are next passed, and *Lake George* entered, 6 miles further, being 32 miles from Lake Huron. This lake or expansion of the river is 9 miles in length and 4 miles broad, affording 12 feet of water over the shoals and terminating at *Church's Landing,* lying opposite *Squirrel Island,* attached to Canada. *Garden River Settlement,* 3 miles, is an Indian town on the Canada side. *Little Lake George* is passed and *Point Aux Pins* reached, 3 miles further. From Little

Lake George to the *Saut Ste. Marie*, passing around the head of Sugar Island, is 8 miles further, being 55 miles from Lake Huron. The *Rapids*, or *Ship Canal*, extend for about one mile, overcoming a fall of 20 feet, when a beautiful stretch of the river is next passed and *Waiska Bay* entered, 6 miles above the rapids; making the St. Mary's River 62 miles in length. The channel forming the boundary line between Canada and the United States is followed by the ascending steamer from the lower end of St. Joseph's Island to Lake Superior, while a more direct passage is afforded for vessels of light draught through *Hay Lake*, lying west of Sugar Island and entering Mud Lake. Nothing can be more charming than a trip over these waters, when sailing to or from the Straits of Mackinac. thus having in view rich and varied lake and river scenery, once the exclusive and favored abode of the red man of the forest, now fast passing away before the march of civilization.

Saut Ste. Marie,* capital of Chippewa Co., Mich., is advantageously situated on St. Mary's River, or Strait, 350 miles N.N.W. of Detroit, and 15 miles from the foot of Lake Superior, in N. lat. 46° 31'. The Rapids at this place, giving the name to the settlements on both sides of the river, have a descent of 20 feet within the distance of a mile, and form the natural limit of navigation. The Ship Canal, however, which has recently been constructed on the American side, obviates this difficulty. Steamers of a large class now pass through the locks into Lake Superior,

* Settled in 1668, by the French.

greatly facilitating trade and commerce. The village on the American side is pleasantly situated near the foot of the rapids, and contains a court-house and jail; a Presbyterian, a Methodist, and a Roman Catholic church; 2 hotels, and 15 or 20 stores and storehouses, besides a few manufacturing establishments, and about 1,200 inhabitants. Many of the inhabitants and Indians in the vicinity are engaged in the fur trade and fisheries, the latter being an important and profitable occupation. Summer visitors flock to this place and the Lake Superior country for health and pleasure. The *Chippewa House,* a well kept hotel on the American side, and one on the Canadian side of the river, both afford good accommodations.

FORT BRADY is an old and important United States military post contiguous to this frontier village, where are barracks for a full garrison of troops. It commands the St. Mary's River and the approach to the mouth of the canal.

SAUT STE. MARIE, C. W., is a scattered settlement, where is located a part of the Hudson Bay Company. Here is a steamboat landing, a hotel, and two or three stores, including the Hudson Bay Company's; and it has from 500 to 600 inhabitants. Indians of the Chippewa tribe reside in the vicinity in considerable numbers, they having the exclusive right to take fish in the waters contiguous to the rapids. They also employ themselves in running the rapids in their frail canoes, when desired by citizens or strangers—this being one of the most exhilarating enjoyments for those fond of aquatic sports. (*See Engraving.*)

SAULT ST. MARIE—FROM AMERICAN SIDE

St. Mary's Falls Ship Canal.

This Canal, which connects the navigation of Lake Superior with the Lower Lakes, is one mile in length, and cost about one million dollars.

It was built in the years 1853, '54, '55, by the Saint Mary's Falls Ship Canal Company, under a contract with commissioners appointed by the authorities of the State of Michigan to secure the building of the canal.

A grant of 750,000 acres of the public land had previously been made by Congress to the State of Michigan, to aid in the construction of this important work.

This grant of 750,000 acres was given to the parties contracting for the building of the canal, provided the work should be completed within two years from the date of the contract.

The work was commenced in the spring of 1853, and completed within the time specified in the contract (*two years!*).

This result was accomplished under many disadvantages, during a very sickly season, and when great difficulty was experienced in obtaining laborers; but the unremitting vigor of those who had the charge of the work secured its completion in the most substantial, permanent, and acceptable manner.

During a great portion of the time there were from 1,200 to 1,600 men employed upon the work, exclusive of the force at the different quarries where the stone was cut and prepared for the locks, beside a large force employed in necessary agencies, getting timber, etc.

The stones for the locks were cut at Anderdon, Canada (near Malden), and at Marblehead, near Sandusky, in Ohio. These were sent in vessels to the work, some twenty-five different sailing vessels being employed in this business.

On the completion of the canal in June, 1855, the governor of the State, the State officers, and the Canal Commissioners proceeded to Saut Ste. Marie for the purpose of inspecting the work. It was accepted, and thereupon, in accordance with the terms of the contract, the State authorities released to the Canal Company and issued patents for the 750,000 acres of land. This was all the remuneration the company received for the work.

The lands were selected during the building of the canal, by agents appointed by the governor of Michigan.

Of the 750,000 acres, 39,000 acres were selected in the iron region of Lake Superior, 147,000 acres in the copper region, and the balance, 564,000 acres, in the Lower Peninsula.

The following figures will give some idea of the magnitude of this work :

Length of canal, 5,548 feet, — 1 mile 301 feet.

Width at top, 115 feet—at water-line, 100 feet—at bottom, 64 feet.

The depth of the canal is 12 feet.

A slope wall on the sides of the canal is 4,000 feet in length.

There are two locks, each 350 feet in length.

Width of locks, 70 feet at top—61½ feet at bottom.

The walls are 25 feet high—10 feet thick at bottom.

Lift of upper lock, 8 feet—lower do., 10 feet ; total lockage, 18 feet.

Lower wharf, 180 feet long; 20 feet wide. Upper wharf, 830 feet long; from 16 to 30 feet wide.

There are 3 pairs of folding gates, each 40 feet wide.

Upper gate, 17 feet high—lower gate, 24 feet 6 inches high.

There are also upper and lower caisson gates, used for shutting off the water from the canal.

The amount of lumber, timber, and iron used in the building of the piers and gates is enormous.

There were 103,437 lbs. of wrought iron used in the gates, and 38,000 lbs. cast iron.

About 8,000 feet of oak timber, etc.

The tolls on the canal are collected by the State—are merely nominal—and only intended to defray the necessary expenses of repairs.

THE ST. MARY'S FALLS SHIP CANAL, Michigan, now forms a navigable communication between Lake Superior and Huron, passing through the St. Mary's River for a distance of about 60 miles.

The first Steamer which passed through the locks was the ILLINOIS, 927 tons, commanded by John Wilson, on her trip through to the upper ports on Lake Superior, June 18, 1855. The Illinois was followed by the Steamer Baltimore, 514 tons; Samuel Ward, 434 tons; and the North Star, 1,100 tons, during the month of June of the same year.

OPENING AND CLOSING OF NAVIGATION, from 1855 to 1862, inclusive.

Date.	First Vessel.			Date.	Last Vessel.		
June 18, 1855	Illinois,	927	tons.	Nov. 23, 1855	Planet,	1,154	tons.
May 4, 1856	Manhattan,	320	"	Nov. 28, 1856	Gen. Taylor,	462	"
May 9, 1857	North Star,	1,100	"	Nov. 30, 1857	Mineral Rock,	555	"
April 18, 1858	Iron City,	600	"	Nov. 20, 1858	Lady Elgin,	1,038	"
May 3, 1859	Lady Elgin,	1,038	"	Nov. 28, 1859	Forester,	384	"
May 11, 1860	Fountain City,	820	"	Nov. 22, 1860	Montgomery,	879	"
May 3, 1861	Michigan,	642	"	Nov. 28, 1861	Gen. Taylor,	462	"
April 27, 1862	City of Cleveland,	788	"	Nov. 27, 1862	Mineral Rock,	555	"
April 28, 1863	Mineral Rock,	555	"				

Average season of navigation, 6¼ months.

Rate of Toll, 6 cents for every registered ton, for every description of vessel.

TABLE OF DISTANCES
From Toronto to Collingwood and Saut Ste. Marie.,

TORONTO TO COLLINGWOOD (*Railroad Route*), 94 miles.

STEAMBOAT ROUTE.

(Collingwood to Saut Ste. Marie, Mich., passing through Georgian Bay and North Channel.)

Ports, etc.	Miles.	Ports, etc.	Miles.
COLLINGWOOD	0	SAUT STE. MARIE.	0
Cape Rich	30	Sugar Island	4
Cabot's Head	80	Garden River Set.	10
Lonely Island	100	*Church's Landing*	14
Cape Smyth	125	Lake George	20
She-ba-wa-nah-ning	145	Nebish Rapids	24
Man-i-tou-wah-ning (25 m.)		St. Joseph Island	25
Little Current,	170	The Narrows	35
Great Manitoulin Is.		Campement D'Ours Is	38
Clapperton Island	190	*Bruce Mines*	50

Ports, etc.	Miles.	Ports, etc.	Miles.
Barrie Island	220	Drummond's Island, Mich.	70
Cockburn Island	255	Cockburn Island, C. W.	85
Drummond's Island, Mich.	270	Barrie Island	120
Bruce Mines, C. W.	290	Clapperton Island	150
St. Joseph Island	296	*Little Current*	
Tampement D'Ours Is.	302	Great Manitoulin Is. }	170
The Narrows	305	Man-i-tou-wah-ning (25 m.)	
Sugar Island, Mich.	315	*She-ba-wa-nah-ning*	195
Nebish Rapids	316	Cape Smyth	215
Lake George	320	Lonely Island	240
Church's Landing	326	Cabot's Head	260
Garden River Set.	330	Cape Rich	310
SALT STE. MARIE	340	COLLINGWOOD	340

STEAMBOAT FARE, $8 50.   USUAL TIME, 36 hours.
Including meals.

NOTE.—Landings in *Italic.*

Collingwood, 94 miles north from Toronto, is most advantageously situated near the head of Nottawassaga Bay, an indentation of Georgian Bay. The town, although commenced in 1854, at the time of the completion of the Ontario, Simcoe, and Huron Railroad, now contains (1861) about 2,000 inhabitants, and is rapidly increasing. The surprising growth is mainly owing to its being the northern terminus of the railway which connects the Georgian Bay with Lake Ontario at Toronto. Great numbers of travellers and emigrants are at this point transferred to steamers or propellers, bound for Mackinac, Green Bay, Chicago, and the Great West, as well as to the Sault Ste. Marie and Lake Superior. Here are a long pier, 800 feet in length; a breakwater, and light-house; several large stores and storehouses; four hotels, and two or three churches in the course of erection.

The steamers leaving Collingwood for Mackinac and Chicago, running along the west shore of Lake Michigan, are of a large class, affording good accommodations for travellers. Steamers run every day to Owen's Sound, 50 miles distant; and weekly to Bruce Mines, the Sant Ste. Marie, and into Lake Superior, affording a delightful steamboat excursion.

Immense quantities of fish are taken in the waters of Nottawassaga Bay, being principally carried to the Toronto market. The whole north shore of the Georgian Bay abounds in white fish, salmon, trout, maskalonge, and other fish of fine quality, affording profitable employment to the Canadians and Indians.

"Some idea of the value and extent of the fishing operations promiscuously pursued in Nottawassaga Bay may be formed from the knowledge that the average daily take exceeds one thousand fish, weighing from forty pounds down to one pound. At this rate, that of the season would not fall short of £40,000. At the mouth of the Nottawassaga River the white fish are netted in perfect shoals throughout the spawning season. Most of the larger kind of trout spawn about the islands upon beds of calcareous rock, over which a shifting drift of sand or gravel passes by the action of the waves, where the water is shallow; and from being exposed to the sun, the temperature of the lake is warmer at these localities than elsewhere. Thither the fishermen resort, and net the fish, vapid and placid as they are, in fabulous amounts."

GEORGIAN BAY.

The deeply romantic character of this pure and lovely body of water is almost unknown to the American public—lying as it does to the northeast of Lake Huron, being entirely within the confines of Canada. The northeast shore is the most romantic and highly interesting, from the fact of there being innumerable islands and islets along the coast, greatly exceeding in number the "Thousand Islands" of the St. Lawrence.

From Penetanguishene, northeast to She-ba-wa-nah-ning, where commences the picturesque body of water known as the *North Channel*, there is one continued succession of enchanting scenery. Here the wild fowl, fur-bearing game, and the finny tribe disport in perfect freedom, being as yet far removed from the busy haunts of civilization.

Georgian Bay is nearly as large as Lake Ontario, while the North Channel, connecting with St. Mary's River on the west, may be said to be as large as Long Island Sound, dotted with a large number of lovely islands, while to the south lies the romantic island of the *Great Manitoulin*, and on the north rises *La Cloche Mountain*—altogether forming the most grand and romantic scenery.

ROUTE FROM COLLINGWOOD, C. W., TO THE SAUT STE. MARIE.

THROUGH GEORGIAN BAY AND NORTH CHANNEL.

This is a new and highly interesting steamboat excursion, brought into notice by the completion of the *Ontario, Simcoe, and Huron Railroad*, extending from Toronto to Collingwood, at the southern extremity of Georgian Bay.

NOTTAWASSAGA BAY, the southern termination of Georgian Bay, is a large expanse of water bounded by Cape Rich on the west, and Christian Island on the east, each being distant about 30 miles from Collingwood. At the south end of the bay lies a small group of islands called the *Hen and Chickens*.

CHRISTIAN ISLAND, lying about 25 miles from Penetanguishene, and 25 miles north-east of Cape Rich, is a large and fertile island, which was early settled by the Jesuits. There are several others passed north of Christian Island, of great beauty, while still farther northwest are encountered innumerable islands and islets, forming labyrinths, and secluded passages and coves as yet almost unknown to the white man, extending westward for upward of one hundred miles.

PENETANGUISHENE, C. W., 50 miles north of Collingwood by steamboat route, situated on a lovely and secure bay, is an old and very important settlement, comprising an Episcopal and Roman Catholic church, two hotels, a custom-house, several stores and storehouses, and has about 500 inhabitants. In the immediate vicinity are a naval and military depot and barracks, established by the British government. The natural beauties of the bay and harbor, combined with the picturesque scenery of the shores, make up a picture of rare beauty. Here may be seen the native Indian, the half-breed, and the Canadian *Voyageur*, with the full-blooded Englishman or Scotchman, forming one community. This place, being near the mouth of the River Severn, and contiguous to the numberless islands of Georgian Bay, is no doubt destined to become a favorite resort for the angler and sportsman, as well as for the invalid and seeker of pleasure.

On leaving *Collingwood* for Bruce Mines and the Saut Ste. Marie, the steamer usually runs direct across Georgian Bay to Lonely Island, passing Cabot's Head to the right, and the passage leading into the broad waters of Lake Huron, which is the route pursued by the steamers in the voyage to Mackinac, Green Bay, and Chicago. During the summer months the trip from Collingwood to Mackinac and Chicago affords a delightful excursion.

OWEN'S SOUND, or SYDENHAM, 50 miles west of Collingwood, although off the direct route to the Saut Ste. Marie, is well worthy of a passing notice. Here is a thriving settlement, surrounded by a fertile section of country, and containing about 2,500 inhabitants. A steamer runs daily from Collingwood to this place, which will, no doubt, soon be reached by railroad.

LONELY ISLAND, situated about 100 miles west of Collingwood and 20 miles east of the Great Manitoulin Islands, is a large body of land mostly covered with a dense forest, and uninhabited, except by a few fishermen, who resort here at certain seasons of the year for the purpose of taking fish of different kinds. The steamer usually passes this island on its north side, steering for *Cape Smyth*, a bold promontory jutting out from the Great Manitoulin, and distant from Lonely Island about 25 miles.

SQUAW ISLAND and PAPOOSE ISLAND are seen on the northeast, while farther inland are the *Fox Islands*, being the commencement on the west of the innumerable islands which abound along the north shore of Georgian Bay.

LA CLOCHE MOUNTAINS, rising about 2,000 feet above the sea, are next seen in the distance, toward the north; these, combined with the wild scenery of the islands and headlands, form a grand panoramic view, enjoyed from the deck of the passing steamer.

SMYTH'S BAY is passed on the west, some eight or ten miles distant. At the head of this bay, on the great Manitoulin Island, are situated a village of Indians, and a Jesuit's mission, called We-qua-me-kong. These aborigines are noted for their industry, raising wheat, corn, oats, and potatoes in large quantities. This part of the island is very fertile, and the climate is healthy.

SHE-BA-WA-NAH-NING, signifying, in the Indian dialect, "*Here is a Channel*," is a most charming spot, 40 miles distant from Lonely Island, hemmed in by mountains on the north and a high rocky island on the south. It is situated on the north side of a narrow channel, about half a mile in length, which has a great depth of water. Here are a convenient steamboat landing, a church, a store, and some ten or twelve dwellings, inhabited by Canadians and half-breeds. Indians assemble here often in considerable numbers, to sell their fish and furs, presenting with their canoes and dogs a very grotesque appearance. One resident at this landing usually attracts much attention—a noble dog, of the color of cream. No sooner does the steamer's bell ring, than this animal rushes to the wharf, sometimes assisting to secure the rope that is thrown ashore;

the next move he makes is to board the vessel, as though he were a custom-house officer; but on one occasion, in his eagerness to get into the kitchen, he fell overboard; nothing daunted, he swam to the shore, and then again boarding the vessel, succeeded in his desire to fill his stomach, showing the instinct which prompts many a biped office-seeker.

On leaving She-ba-wa-nah-ning and proceeding westward, a most beautiful bay is passed, studded with islands; and mountains upwards of 1,000 feet in height, presenting a rocky and sterile appearance, form an appropriate background to the view; thence are passed Badgley and Heywood Islands, the latter lying off Heywood Sound, situated on the north side of the Great Manitoulin.

MAN-I-TOU-WAH-NING, 25 miles northwest of She-ba-wa-nah-ning, is handsomely situated at the head of Heywood Sound. It is an Indian settlement, and also a government agency, being the place annually selected to distribute the Indian annuities.

LITTLE CURRENT, 25 miles west of She-ba-wa-nah-ning, is another interesting landing on the north shore of the Great Manitoulin, opposite La Cloche Island. Here the main channel is narrow, with a current usually running at the rate of five or six knots an hour, being much affected by the winds. The steamer stops at this landing for an hour or upward, receiving a supply of wood, it being furnished by an intelligent Indian or half-breed, who resides at this place with his family. Indians are often seen here in considerable numbers. They are reported to be indolent and harmless, too often neglecting the cultivation of the soil for the more uncertain pursuits of fishing and hunting, although a considerably large clearing is to be seen indifferently cultivated.

CLAPPERTON ISLAND and other islands of less magnitude are passed in the *North Channel*, which is a large body of water

about 120 miles long and 25 miles wide. On the north shore is situated a post of the Hudson Bay Company, which may be seen from .the deck of the passing steamer.

COCKBURN ISLAND, 85 miles west of Little Current lies directly west of the Great Manitoulin, from which it is separated by a narrow channel. It is a large island, somewhat elevated, but uninhabited, except by Indians.

DRUMMOND ISLAND, 15 miles farther westward, belongs to the United States, being attached to the State of Michigan. This is another large body of land, being low, and as yet mostly uninhabited.

The next Island approached before landing at Bruce Mines is ST. JOSEPH ISLAND, being a large and fertile body of land, with some few settlers.

BRUCE MINES VILLAGE, C. W., is situated on the north shore of Lake Huron, or the "North Channel," as it is here called, distant 290 miles from Collingwood, and 50 from the Saut Ste. Marie. Here are a Methodist chapel, a public-house, and a store and storehouse belonging to the Montreal Copper Mining Company, besides extensive buildings used for crushing ore and preparing it for the market; about 75 dwellings and 600 inhabitants. The copper ore, after being crushed by powerful machinery propelled by steam, is put into puddling troughs and washed by water, so as to obtain about 20 per cent. pure copper. In this state it is shipped to the United States and England, bringing about $80 per ton. It then has to go through an extensive smelting process, in order to obtain the pure metal. The mines are situated in the immediate vicinity of the village, there being ten openings or shafts from which the ore is obtained in its crude state. Horse-power is mostly used to elevate the ore; the whims are above ground, attached to which are ropes and buckets. This mine gives employment to about 300 workmen.

The capital stock of the company amounts to $600,000.

The *Wellington Mine*, about one mile distant, is also owned by the Montreal Mining Company, but is leased and worked by an English company. This mine, at the present time, is more productive than the Bruce Mines.

The Lake Superior *Journal* gives the following description of the Bruce Mine, from which is produced a copper ore differing from that which is yielded by other mines of that peninsula.

"Ten years ago this mine was opened, and large sums expended for machinery, which proved useless, but it is now under new management, and promises to yield profitably. Twelve shafts have been opened, one of which has been carried down some 330 feet. Some 200 or 300 men are employed, all from the European mines. Some of the ores are very beautiful to the eye, resembling fine gold. After being taken out of the shaft, they are taken upon a rail-track to the crushing-house, where they are passed between large iron rollers, and sifted till only a fine powder remains; from thence to the 'jigger-works,' where they are shaken in water till much of the earthy matter is washed away, after which it is piled in the yard ready for shipment, having more the appearance of mud than of copper. It is now mostly shipped to Swansea, in Wales, for smelting. Two years since, 1,500 tons were shipped to Baltimore and Buffalo to be smelted."

On resuming the voyage after leaving Bruce Mines, the steamer runs along St. Joseph Island through a beautiful sheet of water, in which are embosomed some few islands near the main shore.

CAMPEMENT D'OURS is an island passed on the left, lying contiguous to St. Joseph Island. Here are encountered several small rocky islands, forming an intricate channel called the "*Narrows*." On some of the islands in this group are found copper ore, and beautiful specimens of moss. The forest-trees, however, are of a dwarfish growth, owing, no doubt, to the scantiness of soil on these rocky islands.

About 10 miles west of the "Narrows," the main channel of the St. Mary's River is reached, forming the boundary between the United States and Canada. A rocky island lies on the Canadian side, which is reserved for government purposes, as it commands the main or ship channel.

SUGAR ISLAND is now reached, which belongs to the United States, and the steamers run a further distance of 25 miles, when the landing at the Saut Ste. Marie is reached, there being settlements on both sides of the river. The British boats usually land on the north side, while the American boats make a landing on the south side of the river, near the mouth of the ship canal.

TRIP FROM COLLINGWOOD TO FORT WILLIAM, C. W.

THE FIRST TRIP OF THE STEAMER RESCUE.

"*To the Editor of the Toronto Globe.*

"SIR: As you have on all occasions taken a prominent part in advocating the opening up of the Hudson's Bay Territory and the North Shore of Lake Superior, I send you a log journal of the first cruise of the Steamer *Rescue*, Captain JAMES DICK, from Collingwood to Fort William. On this trip she fairly maintained her previous reputation; for in a heavy gale of wind on the beam for many hours, between Michipicoten Island and Fort William, she made her 10½ miles per hour, and, during the gale, was steady, and free from any unpleasant motion. We left Collingwood at 10.30 A. M., on the 12th July, 1858, Captain Kennedy in charge of the mails, for Red River. We passed Cabot's Head at 6.30 P. M.; Cove Island light, at 9 P. M. (merely a lantern on the top of the tower, visible about two miles on a clear night); passed between the middle and western Duck Islands at 4 A.M., at easy steam, so as to enter the Missisaga Straits in daylight; at 11.20 A. M., ran alongside the wharf at *Bruce Mines;* landed mails, and wooded. Under the kind supervision of Mr. Davidson we inspected the process of extracting copper ore from the bowels of the earth. We found that it contained 4 per cent. at the mouth of the pit, and 25 per cent. barrelled up in the form of paste. Sometime ago, the Montreal Mining Company (owning the Bruce Mines), leased half their location to the Wellington Mining Company. There are, in consequence, within one mile, separated by a small island, two establishments, forming one considerable town. Arrived at Saut Ste. Marie, Pim's wharf (British side), at 7 P. M.; landed mails, and ran over to the American side for coals. At 6 A. M., on the 14th, entered the ship-canal, paying six cents per ton lockage dues. Mr. Simpson, of the Hudson's Bay Company, very politely sent with us the Captain of their schooner to pilot us through to Pine Point, where we engaged his son-in-law, Alex. Clark, as pilot.

"Passed White Fish Point, Lake Superior, at 10 o'clock A. M., Caribou Island at 4.30 o'clock P. M. This island was so-called, from the circumstance of Captain McHargo, who accompanied Bayfield in his survey, having on one occasion killed 60 Caribous on it. At 6 P. M., we were close to 'Rescue' Harbor, Island of Michipicoten. The harbor at Michipicoten is described by the pilot, who has been 15 years on the lake, as superb, and is so laid down by Bayfield. The island is about 16 miles by 6, covered with spruce,

fir, birch, ash and maple, the latter growing on elevated ground. There are several lakes upon it, full of speckled trout; the bay is full of salmon, trout, and white fish. A schooner was loaded here last season in a very short time with fish in and about the harbor; and the climate is said by old *voyageurs* to be far more pleasant during the winter than at the Saut and other places farther south, being of a drier nature[*] Between the island and the main land is the most sheltered passage, with two excellent harbors on each side, one at Otters' Creek and the other at Michipicoten River and harbor. This latter place is an important port of the Hudson's Bay Company, distant from *Moose Fort,* Hudson's Bay, 300 miles, which has been passed over in canoes in six days. Michipicoten Island is said to contain great mineral deposit—silver, copper, and lead; the Quebec Mining Company have a location here.

At daybreak on Thursday we passed *Slate Island,* and shortly after encountered a dense fog and lay to till 1 P. M. It was two o'clock before we saw land. Passed close to Thunder Cape, a perpendicular rock rising from the water's edge 1,350 feet. Anchored at FORT WILLIAM, situated at the mouth of Kaministoguoi River at 7 P. M., on Thursday 15th, and landed the mail. Owing to a bar and shoal at the mouth of the river, we anchored about a mile from the Fort, early on Friday the 16th. Some of the party went up the river in canoes to the *Jesuit Mission,* about three miles, where they were kindly received by the priest. Capt. Jas. Dick and Mr. McMurrich went fishing to Current River, about five miles to the north, where the speckled trout proved too large and

[*] The romantic and uninhabited harbor on the south side of Michipicoten Island, exceeds in safety, extent, and grandeur any harbor found on the shores of these great lakes. It was visited by the steamer *"Ploughboy,"* with a party of pleasure on board in 1860, lying at anchor all night.

strong for their light rods and tackle, smashing the tops of their rods and tearing away their lines and flies as fast as they were thrown in, and they had to give it up for want of material. One of the trout caught was the largest speckled trout I have seen for some years. There are trout in this stream, and in all the rapid streams between the Saut and Fort William, from 2 lbs. to 6 lbs., and if larger ones are required, at Neepigon River they can be caught from 8 lbs. to 12 lbs. Fancy such a spot, ye disciples of Isaac Walton; speckled trout to be had for the trouble of throwing a fly, within 3½ days of Toronto, weighing from 2 lbs. to 12 lbs. In this vicinity are to be found beautiful specimens of amethyst and other precious stones.

"The gardens at Fort William and at th Jesuit Mission are as forward as those on the north part of the county of Simcoe. The Hudson's Bay Company have a large farm, 50 cows besides horses and sheep, and up the river there are other farms; they raise oats, barley, and all kinds of vegetables, and I see no reason why they cannot raise wheat. Mrs. McIntyre, the wife of the agent, was very polite and kind, and invited us all up to the Fort—gave us supplies of milk and vegetables. By this route their trade is carried on to Red River. Sir George Simpson returned from Red River just before we arrived with two canoes (9 men in each) and left again for the Saut. This bay, Black Thunder, Neepigon Bay, and Pie Island Bay and neighborhood, abound in white fish and trout—10 fish frequently fill a barrel—20 as a general rule; nets should be 5½ to 7½ inch mesh. Our pilot, two years ago, in five weeks, with two men, filled 175 barrels; he was furnished by merchants at the Saut with barrels and salt, and $5 when returned full—the rate this year being about $4. Thirty barrels of white fish were taken at one haul of a seine near Fort William.

We left Fort William at 8 P. M., for *Grand Portage*, passing McKay's Mount of Greenstone, 1,000 feet perpendicular height. La Pate or Pie Island, 850 feet perpendicular; this island is said to abound in lead; hardly a stone can be picked up on the shore without lead in it. On all these islands valuable stones can be picked up, fit for brooches and rings. The channel being very intricate, and the pilot not quite posted up, we lay to till daybreak, and entered *Grand Portage Bay* at 5 A. M. Capt. Kennedy landed here with the mails, purchased a canoe, and was ready to start before we left. A nucleus of a town has already sprung up here on the United States side.

"After giving Capt. Kennedy a hearty shake of the hand all round, we started homeward, at 7.15 A. M., and passed Copper Harbor at 2 P. M., Manitou Light 5.50 P. M., White Fish Point 6.40 A. M.; and on the 18th July entered the Saut Canal at 10.18 A. M. Coaled on the American side, and wooded on Pim's wharf, British side. Landed the mail, and started at 2.45 P. M. Came to the wharf at Bruce Mines at seven P. M.—wooded and left at 7.40—passing through the Mississaga Straits and the channel between the Middle and West Duck. On the 19th passed Cove Island light at 8 A. M.,—Cabot's Head 10 o'clock A. M., and came to the wharf at Collingwood at 6 P. M. Thus making the first trip, including delays and stoppages round Lake Superior, in *seven days and six hours*; distance run, taken from Bayfield's chart, between 1,250 and 1,300 (geographical) miles. The average speed *running time*, being a little over *ten miles per hour*.

"The scenery throughout, and especially that of Superior, is magnificent. And now that the means of communication are afforded to this great and unknown region, in a safe and commodious boat, under the care of a well-known and experienced captain, it must become the favorite route for the tourist in search of health and picturesque scenery."

The Compiler of this volume having, during the summer of 1860, passed over the same route on board the Canadian steamer PLOUGHBOY, can vouch for the accuracy of most of the above described trip, exceeding in rich variety of lake and river scenery any other excursion, of equal extent, on the continent of America.

The shores or mainland, together with virgin islands, are in view for most of the distance, except while crossing the wide waters of Lake Superior,—when all the vastness of the ocean-deep is realized,—you then being surrounded by an unbroken waste of waters.

Distance around Lake Superior.

SAUT STE. MARIE to FORT WILLIAM, C. W., 300 miles; Fort William to Superior City, Wis., 200 miles; Superior City to Saut Ste. Marie (American side), 365 miles—making the grand circuit of Lake Superior, 1,065 miles.

Distances from the Saut. Ste. Marie to Superior City.

Ports, &c.	Distances.	Miles.
SAUT STE. MARIE, Mich.....		00
Point Iroquois.............	15	15
White Fish Point...........	25	40
Point au Sable :...........	50	90
Pictured Rocks.............	20	110
Grand Island...............	10	120
Munising, Mich............	5	125
MARQUETTE, ".............	45	170
Huron Islands.............	45	215
Portage Entry.............	25	240
Hough'·n (Portage Lake, 14 m.)		
Manitou Island, or Keweenaw Point............	60	300
Copper Harbor............	15	315
Agate Harbor............	10	325
Eagle Harbor.............	6	331
Eagle River.............	9	340
ONTONAGON, Mich........	65	405
Porcupine Hills...........	25	430
La Pointe, Wis............	52	482
Bayfield	3	485
Point de Tour.............	10	495
SUPERIOR CITY, Wis........	70	565

GRAND PLEASURE EXCURSION AROUND LAKE SUPERIOR.

On leaving the Ship Canal, at the Saut, the steamer ascends a beautiful stretch of the St. Mary's River for 10 miles before reaching Waiska Bay, being an expansion of the river of about 5 miles. Here the shores assume a bold appearance well worthy the attention of the traveller before launching out on the waters of the broad lake.

IROQUOIS POINT, on the American side, and GROS CAP, on the Canadian side, are next passed, 15 miles from the Saut Ste. Marie. The latter is a bold promontory, rising some 400 or 500 feet above the water, with still higher hills rising in the distance.

TONQUAMENON BAY is next entered, and a scene of grandeur is presented to the view; on the southwest or American shore the land rises to a moderate height, while on the northeast or Canada shore the land rises to mountain height, being elevated from 800 to 1,000 feet, running off far in the distance toward the north.

PARISIEN and other islands, attached to Canada, are passed on the right, the bay being about 25 miles long and as many broad; in fact, forming a part of Lake Superior, whose pure waters are in full view as far as the eye can reach.

GOULAIS BAY, and POINT, another bold headland, lie to the north of Gros Cap, where enters a river of the same name, and are situated on the Canada side. Here are fine fishing-grounds in the bay, while the river abounds in speckled trout, being a favored resort for fishing-parties during pleasant weather.

Lake Superior, by far the largest of the Inland Seas of North America, lying between 46° 30′ and 49° north latitude, and between 84° 30′ and 92° 30′ west longitude, situated at a height of 600 feet above the sea, from which it is distant about 1,500 miles by the course of its outlet and the St. Lawrence River, is 460 miles long from east to west, and 170 miles broad in its widest part, with an average breadth of 85 miles. It is 800 feet in greatest depth, extending 200 feet below the level of the ocean; estimated area, 32,000 square miles. Near two hundred rivers and creeks are said to flow into the lake, the greater part being small streams, and but few navigable, except by canoes, owing to their numerous falls and rapids. It contains several islands, the most important of which are *Isle Royale*, and *The Twelve Apostles*, near its western extremity, and Grand Island, all attached to the United States; Caribou Island, Michipicoten, St. Ignace, Pie, Slate, and other islands attached to Canada.

KEWEENAW POINT is its most remarkable feature, jutting far out into the lake some sixty or seventy miles. On the range of hills running through this point, about 20 miles wide, are found the most valuable *copper mines* in the world. Its good and secure harbors are but few on the south side of the lake, while on the north shore and islands are several perfectly safe harbors, and easy of access. It discharges its surplus waters by the Strait, or River St. Mary, 60 miles long, into Lake Huron, which lies 27 feet below, most of the descent being at the Saut Ste. Marie, where is a Ship Canal three-fourths of a mile in length, with two locks of 10 feet less each, overcoming a descent of 20 feet.

" The early French Jesuit fathers, who first explored and described this great lake, and published an account of it in Paris, in 1636, describe the form of its shores as similar to that of a bended bow, the northern shore being the arc, and the southern shore the cord, while Keweenaw Point, projecting from the southern shore to near the middle of the lake, is the arrow."

This graphic description is illustrated by a map, prepared by them, which displays the geographical position of its shores with as much fidelity as most of the maps of our day, and proves that those early explorers were perfectly familiar with its outline and shores.

" The coast of Lake Superior is mostly formed of rocks of various kinds, and of different geological groups. With the exception of sandy bars at the mouth of some of the rivers and small streams, the whole coast of the lake is rock-bound; and in some places, but more particularly on the north shore, mountain masses of considerable elevation rear themselves from the water's edge, while mural precipices and beetling crags oppose themselves to the surges of this mighty lake, and threaten the unfortunate mariner, who may be caught in a storm upon a lee-shore, with almost inevitable destruction."

" Father of Lakes! thy waters bend
    Beyond the eagle's utmost view,
When, throned in heaven, he sees thee send
    Back to the sky its world of blue,

" Boundless and deep, the forests weave
    Their twilight shade thy borders o'er,
And threatening cliffs, like giants, heave
    Their rugged forms along thy shore."

There are now situated on the American side of Lake Superior twelve lighthouses, viz., on Point Iroquois; White Fish Point; Grand Island; Marquette Harbor; Portage Entry; Manitou Island, near Keweenaw Point; Copper Harbor;

Eagle Harbor; Eagle River; Ontonagon, at mouth Ontonagon River; La Pointe, on Madeline Island; and Minnesota Point, mouth of St. Louis River.

A government survey of the Upper Lakes, including the St. Mary's River and Straits of Mackinac, is being made by a corps of Topographical Surveyors, which when published will furnish accurate charts of these Inland Seas.

WHITE FISH POINT, and LIGHT-HOUSE, 40 miles from the Saut, lies on the southwest or American shore, forming a conspicuous landmark, while *Mamains Point* is seen on the northeast or Canada shore.

On passing *White Fish Point*, where may be seen a number of "sand-dunes," or hills, and a light-house 75 feet in height, the broad waters of Lake Superior are reached. The steamers usually pursue a westerly course toward Grand Island or Marquette, passing *Point au Sable*, 50 miles farther. During clear weather, the steep sandy hills on the south shore, ranging from 300 to 500 feet in height, may be seen from the deck of the steamer.

POINT AU SABLE, 50 miles from White Fish Point, is the first object of interest seen on the south shore, on the upward trip, from the deck of the passing steamer, which usually runs within sight of land, affording views of a continued succession of interesting points and bold headlands.

The PICTURED ROCKS, 20 miles further, or about 110 miles from the Saut, are next passed, presenting a magnificent appearance at certain times of the day, when favorably seen under the rays of a brilliant sun; then the effect is heightened by the constantly changing appearance of these almost enchanted rocks. The steamers occasionally run close in shore, when the weather is favorable, affording a fine opportunity to examine these wonders of nature.

Trip to Lake Superior.

Extract from a Letter, dated
"ST. ILLINOIS, *off Pictured Rocks*, L. S.,
"July, 31, 1862.

"At sunrise this morning, we approached the far-famed *Pictured Rocks* of Lake Superior, and were favored with one of the most grand scenes imaginable. The sun rose clear, reflecting its rays in the waters of the lake, presenting a gorgeous appearance. The *Sail Rock*, and other points of interest, were distinctly visible, while the steamer was running for the ' *Grand Portail*' of the *voyageurs*, the most remarkable feature of this wonder of nature, varying with every cloud effect as seen from the passing vessel.

"The steamer approached cautiously until she had run her bows under the projecting cliff, then came to a stand-still as quietly as though she was lying at a pier or wharf, giving the numerous passengers a fine opportunity to examine the deep recesses of this immense cavern, the floor being covered with clear, transparent water to the depth of 10 or 20 feet. Inside were visible two lesser openings, where a small boat might pass out into the lake on either side—the *portail* being formed at the termination of a projecting cliff—rising about 200 feet above the lake surface.

"The *Pictured Rocks* are thus briefly described by Foster and Whitney, in their geological report:—'They may be described, in general terms, as a series of sand-stone bluffs, extending along the South shore of Lake Superior, for eight or ten miles, and rising, in most places, vertically from the water, without any beach at the base, to a height varying from 50 to 200 feet.'

"Yours, &c.,   J. D."

The *Pictured Rocks*, of which almost fabulous accounts are given by travellers, are one of the wonders of this "Inland Sea." Here are to be seen the *Cascade Falls* and other objects of great interest. The Amphitheatre, Miners' Castle, Chapel, Grand Portal, and Sail Rock, are points of great picturesque beauty, which require to be seen to be justly appreciated.

Extract from FOSTER and WHITNEY'S Report of the Geology of the Lake Superior Land District:

Pictured Rocks.—"The range of cliffs to which the name of the Pictured Rocks has been given, may be regarded as among the most striking and beautiful features of the scenery of the Northwest, and are well worthy the attention of the artist, the lover of the grand and beautiful, and the observer of geological phenomena.

"Although occasionally visited by travellers, a full and accurate description of this extraordinary locality has not as yet been communicated to the public.*

"The *Pictured Rocks* may be described, in general terms, as a series of sandstone bluffs extending along the shore of Lake Superior for about five miles, and rising, in most places, vertically from the water, without any beach at the base, to a height varying from fifty to nearly two hundred

feet. Were they simply a line of cliffs, they might not, so far as relates to height or extent, be worthy of a rank among great natural curiosities, although such an assemblage of rocky strata, washed by the waves of the great lake, would not, under any circumstances, be destitute of grandeur. To the voyager coasting along their base in his frail canoe they would, at all times, be an object of dread; the recoil of the surf, the rockbound coast, affording for miles no place of refuge; the lowering sky, the rising wind; all these would excite his apprehension, and induce him to ply a vigorous oar until the dreaded wall was passed. But in the Pictured Rocks there are two features which communicate to the scenery a wonderful and almost unique character. These are, first, the curious manner in which the cliffs have been excavated and worn away by the action of the lake, which for centuries has dashed an ocean-like surf against their base; and, second, the equally curious manner in which large portions of the surface have been colored by bands of brilliant hues.

"It is from the latter circumstance that the name by which these cliffs are known to the American traveller is derived; while that applied to them by the French *voyageurs* ('Les Portails'*) is derived from the former, and by far the most striking peculiarity.

"The term *Pictured Rocks* has been in use for a great length of time, but when it was first applied we have been unable to discover.

"The Indian name applied to these cliffs, according to our *voyageurs*, is *Schkuee-archibi-kung*, or 'The end of the rocks,'

---

* Schoolcraft has undertaken to describe this range of cliffs, and illustrate the scenery. The sketches do not appear to have been made on the spot, or finished by one who was acquainted with the scenery, as they bear no resemblance, so far as we observed, to any of the prominent features of the Pictured Rocks.

"It is a matter of surprise that, so far as we know, none of our artists have visited this region, and given to the world representations of scenery so striking, and so different from any which can be found elsewhere. We can hardly conceive of any thing more worthy of the artist's pencil; and if the tide of pleasure-travel should once be turned in this direction, it seems not unreasonable to suppose that a fashionable hotel may yet be built under the shade of the pine groves near the Chapel, and a trip thither become as common as one to Niagara now is."

---

* Le Portail is a French term, signifying the principal entrance of a church or a portal, and this name was given to the Pictured Rocks by the *voyageurs*, evidently in allusion to the arched entrances which constitute the most characteristic feature. Le Grand Portail is the great archway, or Grand Portal.

which seems to refer to the fact that, in descending the lake, after having passed them, no more rocks are seen along the shore. Our *voyageurs* had many legends to relate of the pranks of the *Menni-boujou* in these caverns, and in answer to our inquiries seemed disposed to fabricate stories without end of the achievements of this Indian deity.

"We will describe the most interesting points in the series, proceeding from west to east. On leaving Grand Island harbor,* high cliffs are seen to the east, which form the commencement of the series of rocky promontories, which rise vertically from the water to the height of from one hundred to one hundred and twenty-five feet, covered with a dense canopy of foliage. Occasionally a small cascade may be seen falling from the verge to the base in an unbroken curve, or gliding down the inclined face of the cliff in a sheet of white foam. The rocks at this point begin to assume fantastic shapes; but it is not until having reached Miners' River that their striking peculiarities are observed.

* The traveller desirous of visiting this scene should take advantage of one of the steamers or propellers which navigate the lake and land at Grand Island, from which he can proceed to make the tour of the interesting points in a small boat. The large vessels on the lake do not approach sufficiently near the cliffs to allow the traveller to gather more than a general idea of their position and outlines. To be able to appreciate and understand their extraordinary character, it is indispensable to coast along in close proximity to the cliffs and pass beneath the Grand Portal, which is only accessible from the lake, and to land and enter within the precincts of the Chapel. At Grand Island, boats, men, and provisions may be procured. The traveller should lay in a good supply, if it is intended to be absent long enough to make a thorough examination of the whole series. In fact, an old voyager will not readily trust himself to the mercy of the winds and waves of the lake without them, as he may not unfrequently, however auspicious the weather when starting, find himself weather-bound for days together. It is possible, however, in one day, to start from Grand Island, see the most interesting points, and return. The distance from William's to the Chapel—the farthest point of interest—is about fifteen miles.

Here the coast makes an abrupt turn to the eastward, and just at the point where the rocks break off and the friendly sand-beach begins, is seen one of the grandest works of nature in her rock-built architecture. We gave it the name of 'Miners' Castle,' from its singular resemblance to the turreted entrance and arched portal of some old castle—for instance, that of Dumbarton. The height of the advancing mass, in which the form of the Gothic gateway may be recognized, is about seventy feet, while that of the main wall forming the background is about one hundred and forty. The appearance of the openings at the base changes rapidly with each change in the position of the spectator. On taking a position a little farther to the right of that occupied by the sketcher, the central opening appears more distinctly flanked on either side by two lateral passages, making the resemblance to an artificial work still more striking.

"A little farther east, Miners' River enters the lake close under the brow of the cliff, which here sinks down and gives place to a sand-bank nearly a third of a mile in extent. The river is so narrow that it requires no little skill on the part of the voyager to enter its mouth when a heavy sea is rolling in from the north. On the right bank, a sandy drift plain, covered with Norway and Banksian pine, spreads out, affording good camping-ground —the only place of refuge to the voyager until he reaches Chapel River, five miles distant, if we except a small sand-beach about midway between the two points, where, in case of necessity, a boat may be beached.

"Beyond the sand-beach at Miners River the cliffs attain an altitude of one hundred and seventy-three feet, and maintain a nearly uniform height for a considerable distance. Here one of those cascades of which we have before spoken is seen foaming down the rock.

"The cliffs do not form straight lines, but rather arcs of circles, the space between the projecting points having been worn out in symmetrical curves, some of which are of large dimensions. To one of the grandest and most regularly formed we gave the name of 'The Amphitheatre.' Looking to the west, another projecting point—its base worn into cave-like forms—and a portion of the concave surface of the intervening space are seen.

" It is in this portion of the series that the phenomena of colors are most beautifully and conspicuously displayed. These cannot be illustrated by a mere crayon sketch, but would require, to reproduce the natural effect, an elaborate drawing on a large scale, in which the various combinations of color should be carefully represented. These colors do not by any means cover the whole surface of the cliff even where they are most conspicuously displayed, but are confined to certain portions of the cliffs in the vicinity of the Amphitheatre; the great mass of the surface presenting the natural light-yellow or raw sienna color of the rock. The colors are also limited in their vertical range, rarely extending more than thirty or forty feet above the water, or a quarter or a third of the vertical height of the cliff. The prevailing tints consist of deep-brown, yellow, and gray—burnt sienna and French gray predominating.

" There are also bright blues and greens, though less frequent. All of the tints are fresh, brilliant, and distinct, and harmonize admirably with one another, which, taken in connection with the grandeur of the arched and caverned surfaces on which they are laid, and the deep and pure green of the water which heaves and swells at the base, and the rich foliage which waves above, produce an effect truly wonderful.

" They are not scattered indiscriminately over the surface of the rock, but are arranged in vertical and parallel bands, extending to the water's edge. The mode of their production is undoubtedly as follows: Between the bands or strata of thick-bedded sandstone there are thin seams of shaly materials, which are more or less charged with the metallic oxides, iron largely predominating, with here and there a trace of copper. As the surface-water permeates through the porous strata it comes in contact with these shaly bands, and, oozing out from the exposed edges, trickles down the face of the cliffs, and leaves behind a sediment, colored according to the oxide which is contained in the band in which it originated. It cannot, however, be denied that there are some peculiarities which it is difficult to explain by any hypothesis.

" On first examining the Pictured Rocks, we were forcibly struck with the brilliancy and beauty of the colors, and wondered why some of our predecessors, in their descriptions, had hardly adverted to what we regarded as their most characteristic feature. At a subsequent visit we were surprised to find that the effect of the colors was much less striking than before; they seemed faded out, leaving only traces of their former brilliancy, so that the traveller might regard this as an unimportant feature in the scenery. It is difficult to account for this change, but it may be due to the dryness or humidity of the season. If the colors are produced by the percolation of the water through the strata, taking up and depositing the colored sediments, as before suggested, it is evident that a long period of drought would cut off the supply of moisture, and the colors, being no longer renewed, would fade, and finally disappear. This explanation seems reasonable, for at the time of our second visit the beds of the streams on the summit of the table-land were dry.

" It is a curious fact, that the colors are so firmly attached to the surface that they are very little affected by rains or the dashing of the surf, since they were, in numerous instances. observed extending in all their freshness to the very water's edge.

"Proceeding to the eastward of the Amphitheatre, we find the cliffs scooped out into caverns and grotesque openings, of the most striking and beautiful variety of forms. In some places huge blocks of sandstone have become dislodged and accumulated at the base of the cliff, where they are ground up and the fragments borne away by the ceaseless action of the surge.

"To a striking group of detached blocks the name of 'Sail Rock' has been given, from its striking resemblance to the jib and mainsail of a sloop when spread—so much so that when viewed from a distance, with a full glare of light upon it, while the cliff in the rear is left in the shade, the illusion is perfect. The height of the block is about forty feet.

"Masses of rock are frequently dislodged from the cliff, if we may judge from the freshness of the fracture and the appearance of the trees involved in the descent. The rapidity with which this undermining process is carried on, at many points, will be readily appreciated when we consider that the cliffs do not form a single unbroken line of wall; but, on the contrary, they present numerous salient angles to the full force of the waves. A projecting corner is undermined until the superincumbent weight becomes too great, the overhanging mass cracks, and, aided perhaps by the power of frost, gradually becomes loosened and finally topples with a crash into the lake.

"The same general arched and broken line of cliffs borders the coast for a mile to the eastward of Sail Rock, where the most imposing feature in the series is reached. This is the Grand Portal—*Le Grand Portail* of the *voyageurs.* The general disposition of the arched openings which traverse this great quadrilateral mass may, perhaps, be made intelligible without the aid of a ground-plan. The main body of the structure consists of a vast mass of a rectilinear shape, projecting out into the lake about six hundred feet, and presenting a front of three hundred or four hundred feet, and rising to a height of about two hundred feet. An entrance has been excavated from one side to the other, opening out into large vaulted passages which communicate with the great dome, some three hundred feet from the front of the cliff. The Grand Portal. which opens out on the lake, is of magnificent dimensions, being about one hundred feet in height, and one hundred and sixty-eight feet broad at the water-level. The distance from the verge of the cliff over the arch to the water is one hundred and thirty-three feet, leaving thirty-three feet for the thickness of the rock above the arch itself. The extreme height of the cliff is about fifty feet more, making in all one hundred and eighty-three feet.

"It is impossible, by any arrangement of words, or by any combination of colors, to convey an adequate idea of this wonderful scene. The vast dimensions of the cavern, the vaulted passages, the varied effects of the light, as it streams through the great arch and falls on the different objects, the deep emerald green of the water, the unvarying swell of the lake, keeping up a succession of musical echoes, the reverberations of one's own voice coming back with startling effect, all these must be seen, and heard, and felt, to be fully appreciated.

"Beyond the Grand Portal the cliffs gradually diminish in height, and the general trend of the coast is more to the southeast; hence the rock, being less exposed to the force of the waves, bears fewer marks of their destructive action. The entrance to Chapel River is at the most easterly extremity of a sandy beach which extends for a quarter of a

THE CHAPEL.—PICTURED ROCKS.

mile, and affords a convenient landing-place, while the drift-terrace, elevated about thirty feet above the lake-level, being an open pine plain, affords excellent camping-ground, and is the most central and convenient spot for the traveller to pitch his tent, while he examines the most interesting localities in the series which occur in this vicinity—to wit, the Grand Portal and the Chapel. (*See Engraving*.)

"The Chapel—*La Chapelle* of the *voyageurs*—if not the grandest, is among the most grotesque of Nature's architecture here displayed. Unlike the excavations before described, which occur at the water's edge, this has been made in the rock, at a height of thirty or forty feet above the lake. The interior consists of a vaulted apartment, which has not inaptly received the name it bears. An arched roof of sandstone, from ten to twenty feet in thickness, rests on four gigantic columns of rock, so as to leave a vaulted apartment of irregular shape, about forty feet in diameter, and about the same in height. The columns consist of finely stratified rock, and have been worn into curious shapes. At the base of one of them an arched cavity or niche has been cut. to which access is had by a flight of steps formed by the projecting strata. The disposition of the whole is such as to resemble very much the pulpit of a church; since there is overhead an arched canopy, and in front an opening out toward the vaulted interior of the chapel, with a flat tabular mass in front, rising to a convenient height for a desk, while on the right is an isolated block, which not inaptly represents an altar; so that if the whole had been adapted expressly for a place of worship, and fashioned by the hand of man, it could hardly have been arranged more appropriately. It is

8

hardly possible to describe the singular and unique effect of this extraordinary structure; it is truly a temple of nature— 'a house not made with hands.'

"On the west side, and in close proximity, Chapel River enters the lake, precipitating itself over a rocky ledge ten or fifteen feet in height.*

"It is surprising to see how little the action of the stream has worn away the rocks which form its bed. There appears to have been hardly any recession of the cascade, and the rocky bed has been excavated only a foot or two since the stream assumed its present direction.

"It seems therefore impossible that the river could have had any influence in excavating the Chapel itself, but its excavation must be referred to a period when the waters of the lake stood at a higher level.

"Near the Grand Portal the cliffs are covered, in places, with an efflorescence of sulphate of lime, in delicate crystallizations; this substance not only incrusts the walls, but is found deposited on the moss which lines them, forming singular and interesting specimens, which however cannot be transported without losing their beauty.

"At the same place we found numerous traces of organic life in the form of obscure fucoidal markings, which seem to be the impressions of plants, similar to those described by Prof. Hall as occurring in the Potsdam sandstone of New York. These were first noticed at this place by Dr. Locke, in 1847."

* "At this fall, according to immemorial usage among the *coyageurs* in ascending the lake, the *mangeurs de lard*, who make their first trip, receive baptism; which consists in giving them a severe ducking—a ceremony somewhat similar to that practised on green-horns when crossing the line.

Lake Superior Region.

The following verses were written by J. G. WHITTIER, on receiving an *eagle's quill,* when on a visit to Lake Superior in 1846.

THE SEER.

I hear the far-off voyager's horn,
  I see the Yankee's trail—
His foot on every mountain pass,
  On every stream his sail.

He's whistling round St. Mary's Falls,
  Upon his loaded train ;
He's leaving on the Pictured Rocks
  His fresh tobacco stain.

I see the mattock in the mine,
  The axe-stroke in the dell,
The clamor from the Indian lodge,
  The Jesuit's chapel bell!

I see the swarthy trappers come
  From Mississippi's Springs ;
And war-chiefs with their painted brows,
  And crests of eagle wings.

Behind the scared squaw's birch canoe,
  The steamer smokes and raves ;
And city lots are staked for sale
  Above old Indian graves.

By forest, lake and water-fall,
  I see the peddler's show ;
The mighty mingling with the mean,
  The lofty with the low.

I hear the tread of pioneers
  Of nations yet to be ;
The first low wash of waves where soon
  Shall roll a human sea.

The rudiments of empire here
  Are plastic yet and warm ;
The chaos of a mighty world
  Is rounding into form !

Each rude and jostling fragment soon
  Its fitting place shall find—
The raw materials of a state,
  Its muscle and its mind !

And, westering still, the star which leads
  The new world in its train,
Has tipped with fire the icy spears
  Of many a mountain chain.

GRAND ISLAND, 125 miles distant from the Saut, is about 10 miles long and 5 wide, lying close in to the south shore. This is a wild and romantic island ; the cliffs of sandstone, irregular and broken into by the waves, form picturesque caverns, pillars, and arches of immense dimensions. There are several romantic bays and inlets protected from storms, which are frequent on this great lake, where the brook trout of a large size can be caught in quantities. The forests also afford a delightful retreat, while all nature seems hushed—save by the moaning winds and billowy surges of the surrounding waters.

A few families reside on the south shore, facing the mainland, where is a clearing of considerable extent. The main-shore in full sight, and the Pictured Rocks, visible from its eastern shore, altogether add a charm to this truly Grand Island, unsurpassed by no other spot in this interesting region.

MUNISING, formerly called Grand Island City, lies on the south side of Grand Island Bay, here about 3 miles in width. Here is a steamboat wharf and hotel, together with a few dwellings, being, no doubt, destined to become a favorite place of resort, as from this place the Pictured Rocks can be easily reached by canoes or small boats during calm weather. Trout fishing is also good in Ann's River, which enters Grand Island Bay, and in Miner's River, near the Pictured Rocks.

The bay or harbor is capacious, deep, and easy of access from the east or west, being 6 miles in length by from 2 to 4 in width, with a depth of water of 100 feet and upwards. It is perfectly landlocked by hills rising from 100 to 300 feet high, and capacious enough to contain the entire fleet of the lakes.

It is proposed to construct a railroad from this harbor to the head of Big Bay de Noc, the most northern arm of Green Bay, only 40 miles distant.

MINER'S POINT, a most remarkable headland, lies 6 miles east of Munising, at the mouth of a small stream of the same name.

The action of the waters has here disintegrated portions of the sand-stone formation, forming romantic caverns and grottoes where the waters of the lake penetrate, making strange music in the subterranean passages.

MONUMENT ROCK.

MONUMENT ROCK, about one mile west of Miner's Point, is another strange freak of nature, being an upright column standing in full view, near the water's edge, elevated some 80 or 100 feet above the lake. (*See Engraving.*) All these points can easily be reached from Munising by a sail or row boat, during calm weather.

Remarkable Phenomena on Lake Superior.

The sudden and singular changes of the weather on Lake Superior, in connection with its healthy influence, during the summer and fall months, present one of the phenomena of nature which seems almost unaccountable. The sun frequently rises clear and cloudless, giving indications of continued sunshine, when suddenly the sky becomes overcast with white, fleecy clouds, scudding low and giving out a chilly atmosphere, not unfrequently accompanied with rain,—the clouds as suddenly disappear, and a pleasant afternoon usually follows, with light winds. This influence, causing a fluctuation of several degrees of the thermometer, seems to have an injurious effect on most kinds of fruit and vegetables requiring a warm sun throughout the day in order to arrive at maturity; the country a few miles inland, however, being less subject to these frequent changes.

On the 6th of August, 1860, there occurred a remarkable phenomenon, as witnessed on Grand Island Bay, near the Pictured Rocks—Lake Superior being here about 170 miles wide. During the forenoon of a pleasant summer's day, the water was observed suddenly to fall some three or four feet perpendicularly on the south shore, then rise in about half an hour, as suddenly again to recede and rise several times; exposing the bed of the lake for a considerable distance where the water was shallow, affording a fine opportunity to collect pebbles of different hues, and precious stones.

At noon the wind blew moderately from the southward, while the thermometer ranged at about 74° Fahr. This apparently calm and pleasant weather was taken advantage of by a party of pleasure to cross the bay in a sail-boat from Munising to Grand Island, 3 miles distant, affording a delightful excursion. On looking eastward at about 4 o'clock, P. M., a dense fog or low cloud was seen rapidly to enter the east channel of the bay, from the northward, rolling on in majestic grandeur, and presenting apparently the smoke caused by the discharge of a park of artillery, obscuring every object in the far distance, while the headlands within one or two miles were distinctly visible. As it approached, the thermometer fell several degrees, and rain followed. attended with lightning and thunder. Soon, however, the wind lulled. or entirely ceased, while the rain poured down in torrents. The mist or fog seemed mostly to ascend as it passed over the high lands on the main land, and assumed the appearance of clouds, while portions remained, in low and wet localities, above the forest-trees, —presenting altogether a most magnificent appearance. The rain-storm and cloud effect, after continuing some two hours, as suddenly ceased, followed by a splendid rainbow,—being the harbinger of a pleasant evening and calm weather for a time.

Mackenzie, who wrote in 1789, relates a very similar phenomenon, which occurred at Grand Portage, on Lake Superior, and for which no obvious cause could be assigned. He says: "The water withdrew, leaving the ground dry which had never before been visible, the fall being equal to four perpendicular feet, and rushing back with great velocity above the common mark. It continued thus rising and falling for several hours, gradually decreasing until it stopped at its usual height."

To the mariner these sudden storms and fluctuations, accompanied by fog, are attended with much danger, more particularly if near the land, when the sun and all objects in sight suddenly disappear as if in darkest night, the terrific noise of the waves and wind alone being heard. When followed by snow the danger is still more increased, frequently causing the most disastrous shipwrecks. In this high latitude a perfect calm seldom continues but for a short time; the wind will occasionally lull, when fitful gusts disturb the waters, to be followed by a breeze or storm from some quarter of the compass.

On examining the meteorological record kept at Fort Mackinac, about 100 miles distant in a southeast direction from Grand Island, it was found that the thermometer ranged at 78° Fahrenheit at 2 P. M. on the above day; the wind being from the south. At 7 P. M. a heavy rain and thunder storm commenced, which lasted two hours, the same as on Lake Superior, terminating with a gorgeous sunset view, exceeded only by the magnificent aurora, which frequently illuminates the northern heavens in this high latitude, or the beautiful mirage of mid-day, which reflects with remarkable distinctness the invisible landscape, and vessels floating on the bosom of this vast inland sea.

How far the receding of the waters had to do with the above coming storm, must be left to conjecture or further investigation—no doubt, however, it caused a displacement of water at some remote parts of the lake, which was almost immediately felt at other and far distant points. So with the vapory clouds which suddenly rise over Lake Superior; they, no doubt, being caused by cold currents of air from the higher regions or northwest, passing over warmer portions along the south shore, when immediately a mist or fog is created, which ascends in the form of clouds into the upper regions; not, however, at first very far above the lake level—thus giving out the cold influence above referred to as peculiar to the south shore of the lake when the northwest winds prevail: this cold influence being most probably wafted far to the east and southward, producing, no doubt, an effect on the weather along the Atlantic coast several hundred miles to the southeast. The northwest winds which mostly prevail in the States of New York and Pennsylvania have a modified character, similar to the winds from the same quarter passing over the upper lakes of North America—affording a cool and bracing influence on the human system.

Another remarkable feature in the climate of Lake Superior, is its healthy and invigorating influence on residents and invalids suffering from incipient pulmonary and throat complaints—the sudden changes of hot and cold, or wet weather, seem to brace the constitution, without producing any other injurious effects than rheumatism, when too much exposure is endured.

While the balmy southern clime too often disappoints the invalid, this northern climate, its influence extending westward toward the Rocky Mountains, seems to give strength to the respiratory and digestive organs—thereby often effecting most miraculous and permanent cures,

without the aid of medicine, other than that afforded by nature—pure air and water. The intense colds of winter are here represented as being far more endurable than in more southern latitudes, along the Atlantic coast, where damp northeast storms prevail.

In *Foster and Whitney's Report* on the Geology of Lake Superior, the phenomena of these fluctuations are elaborately discussed; and, for the most part, they are found to be the premonition of an approaching gale. They remark, that the earth may be regarded as surrounded by two oceans—one aërial, the other liquid. By the laws which regulate two fluids thus relatively situated, a local disturbance in the one would produce a corresponding disturbance in the other. Every rise or fall of one-twentieth of an inch in the mercurial column, would be attended with an elevation or depression in the surface of the water equal to one inch. A sudden change of the atmospheric pressure over a large body of water would cause a perpendicular rise or fall, in the manner of waves, greater than the mere weight itself, which would propagate themselves in a series of undulations from the centre of disturbance. These undulations result from an unusual disturbance of the atmosphere occurring around the margin of the storm, and its effects are perceived before the storm actually breaks.

Rise and Fall of the Waters of Lake Superior.

From a series of careful observations continued through a period of eight years, from 1854 to 1862, by Dr. G. H. Blaker, of Marquette, L. S., it has been found that the annual rise and fall of the surface of Lake Superior ranges between 20 and 28 inches. From the first of May, when the snow begins to melt freely, until the first of September, the surface of the lake level continues to rise constantly, about six inches a month, until it gains, on an average, two feet by the middle of August; —and by the first of September it begins to fall, and so continues through the winter, until about the middle of April. The permanent rise, however, was found to have been about *two inches* more than the fall for the first six years, from 1854 to end of 1859, thus making a total rise of some 12 inches in the lake level at the latter period.

During the years 1860 and 1861, the waters of Lake Superior fell about two inches annually, making a fall of four inches since 1859, at which period they were at their *highest point.*

During the winter of 1861-'62, there fell at Marquette only *four* feet and two inches of snow, being about one-quarter the usual amount,—and for the spring months of 1862 there fell only five inches of moisture, being about one-half the usual quantity. These well-authenticated and singular facts, continued to July, 1862, show that the waters of Lake Superior at Marquette are twenty inches lower than they were in 1861—thus showing an unusual depression in the waters of this great inland sea.

When these interesting observations shall have been extended over a longer period and at different stations, they will, no doubt, solve the mystery which has heretofore involved the annual and periodical rise and fall of these great waters in obscurity.

A careful survey of the great lakes by a corps of engineers attached to the Topographical Bureau is now nearly completed, which will give meteorological results and tidal observations of the greatest importance to the mariner, agriculturist, and intelligent traveller.

Marquette, the county seat of Marquette county, and a port of entry, is advantageously situated on the Bay of Marquette, in N. lat. 46° 32', W. long. 87° 41'. The harbor is safe and commodious, being fully protected from all but northeast winds, when vessels are obliged to anchor in the bay for safety. The settlement of Marquette was commenced in July, 1849, and incorporated as a village in June, 1859. It now contains a court-house and jail; 1 Episcopal, 1 Methodist, 1 Presbyterian, 1 Baptist, and 1 Roman Catholic church; 4 public-houses, the *Marquette House* and *Tremont House* being the most frequented by strangers; 2 printing-offices; 15 or 20 stores and storehouses; besides a large number of machine-shops of different kinds. Population in 1860, 1,665.

This flourishing town is identified with the iron-ore beds in the vicinity, being some 12 or 18 miles distant, situated on an elevated ridge being known as the *Iron Mountain.* Here are now three ore-beds extensively and profitably worked, being owned by the Jackson Iron Company, the Cleveland Iron Mining Company, and the Lake Superior Iron Company; each of the above companies have separate docks, from which the ore is shipped to the Eastern markets. A railroad extends from Marquette to the Lake Superior mine, 18 miles, affording ample means for the transportation of iron ore to the place of shipment. The Pioneer Iron Works, situated near the Jackson Iron Mountain, is a large blast furnace giving employment to about 150 workmen. At Collinsville, 3 miles from Marquette, is also a blast furnace employing about 100 hands. At Forrestville, situated on Dead River, is another blast furnace.

The *Northern Iron Company*, situated at Chocolat, 3 miles distant from Marquette, are new works of an extensive character, being largely engaged in the manufacture of pig-iron. In the village are two iron foundries for the manufacture of railroad car-wheels and other castings.

The iron business now gives employment to above 100 sail-vessels, besides several propellers. Steamers of a large class, during the season of navigation, which usually lasts six months, arrive and depart almost daily for Detroit, Cleveland, Milwaukie, and Chicago, carrying freight and passengers.

Carp and Dead rivers both flow into Lake Superior near Marquette, on each side of which there are rapids and falls of great beauty, affording good water-power. Chocolate River also flows into the lake some two or three miles east of Marquette, but through a different geological formation.

The small streams in the vicinity abound in speckled trout, while the lake is at most seasons of the year alive with white-fish, and the Mackinac trout of large dimensions, weighing from 5 to 50 lbs. The climate of Marquette and its vicinity is celebrated for its purity and healthy influence, being the favorite resort of invalids and seekers of pleasure.

NEGAUNEE is a new and thriving settlement, situated on the line of the railroad, 12 miles from Marquette, and in the immediate vicinity of the Iron Mountain. Here is a population of about 1,000 souls, being mostly engaged in working at the mines.

Trout Fishing.

Extract from a MARQUETTE *Paper of July,* 1862.

"The chances for taking trout in our streams and lakes this season appear to be better than ever. Every day we see parties of our citizens or visitors starting out, armed and equipped with all the implements necessary and convenient for the sport, and returning with satisfied countenances and fine strings of 'speckled trout.'

"A considerable quantity have been taken about the rocky points that extend into the bay and lake, while along the Carp, Dead River and smaller streams, the sportsman meets with excellent success.

"To those fond of taking long fishing excursions, and enjoying the luxury of 'camping out,' this country offers extra inducements. Various places, both above and below us, on the lake shore, which are easily reached by sail-boat, are frequented by lovers of sport, who always return with a good supply of trout.

"Back in the country are the Esconawba River and Lake Michigaunie, both of which are within a comfortable day's journey from the terminus of the railroad. The scenery around this lake and along the river is delightful, which, with the abundance of fish to be taken there, well repays the trouble of the excursion."

Bay de Noquet and Marquette Railroad.

This road was commenced in 1853, as a private company, by the late Herman B. Ely and his associates, and chartered in 1855, under the title of the *Iron Mountain Railroad;* finished in 1857 to the Lake Superior Iron Mine, 17 miles distant from Marquette; passing the *Jackson Mine,* 14 miles, and the *Cleveland Mine,* 16 miles. In 1859 it was consolidated with the Bay de Noquet and Marquette Railroad, which will be extended to the head of Little Bay de Noquet, situated on the north end of Green Bay, a total distance of 70 miles. When finished it will form a direct route, by means of railroad and steamers, to all the ports on Green Bay and Lake Michigan. This road has a land grant from government of six sections of timbered land for every mile constructed, amounting to 420 sections of 610 acres each, or 268,800 acres, valued at $672,000.

This is a well-constructed road as far as finished, having an ascending grade for twelve miles, overcoming an elevation of 850 feet before reaching the Iron Mines, thus facilitating the transportation of iron ore to the port of shipment, where extensive piers are constructed for loading of vessels engaged in this growing and important trade. The amount carried over the road in 1860, averaged about 1,500 tons daily, during the season of navigation. It has at the present time (1862) in use four first-class locomotives and 350 freight-cars, with a carrying capacity of 2,500 tons a day.

On leaving Marquette the coast tends north-westward, passing *Presque Isle* and other bold headlands, the coast here presenting a rocky, iron-bound appearance for many miles, with high hills in the distance, being elevated from 800 to 1,000 feet above the waters of the lake.

GRANITE ISLAND, 15 miles north of Marquette, is passed on the right, having on one side two vertical walls of trap, 20 feet high and 12 apart, forming a secure boat harbor. On the mainland opposite is seen *Granite Point*, rising from 120 to 130 feet. Due North from the above island lies *Stanard's Rock*, a most dangerous projection discovered by Captain Stanard in 1835, while in the employ of the American Fur Company, sailing the schooner John Jacob Astor. The rock may be seen on the direct route of steamers from Marquette to Manitou Island or Copper Harbor.

The HURON ISLES, lying about 20 miles east of Portage Entry, numbering five or six rocky islands or islets, form a most picturesque group, covered in part with a stunted growth of trees.

HURON BAY and POINT ABBEYE are next passed, and the upward bound steamer enters a large expanse of water called L'Ance, or Keweenaw Bay, extending far inland.

L'ANCE is an excellent harbor where is a small settlement, situated at the head of Keweenaw Bay. A short distance north are located a Roman Catholic and Methodist mission-house and church. The Catholic being on the west shore of the Bay, and the Methodist on the east, both are surrounded by Indian tribes and settlements. This locality, at no distant day, must become an important point, being favorably situated between the iron and copper regions of Lake Superior.

PORTAGE ENTRY, 70 miles above Marquette, is an important port of entry, here being the mouth of the outlet to Portage Lake, where there stands a light-house to guide the mariner.

The land here is low and the shore uninteresting, except being lined with variegated sandstone, worked into almost every variety of shape by the action of water.

The *Entry* and *Lake* is an extensive and beautiful sheet of water, extending to within half a mile of the entire breadth of the peninsula of Keweenaw Point, in the county of Houghton. It receives a number of small streams, draining the rich copper region of Lake Superior. No portion of the south shore of Lake Superior exceeds this lake and its vicinity as a resort for invalids.

In the immediate vicinity of the lake are found rich deposits of copper, yielding great returns to the miner and capitalist.

HOUGHTON, the county seat of Houghton county, Michigan, and a port of entry, is situated on the south side of Portage Lake, 14 miles from Portage Entry, where its waters commingle with Lake Superior. The harbor is land-locked, being protected by high hills on both sides. The settlement of Houghton was commenced in 1854, and incorporated as a village in 1861. It now contains a court-house and jail; 1 Episcopal, 1 Methodist, and 1 Roman Catholic church; 5 public-houses, the *Douglass House* being a large and well-kept hotel; 10 stores, and several warehouses; 2 steam saw-mills, 2 breweries, and 2 large stamp-mills using steam power. The population of the town is estimated at 3,000, being mostly engaged in mining operations, while the general trade and lumbering afford profitable employment to those engaged in the latter pursuits. This new and flourishing town, lying on a side-hill rising 300 or 400 feet, is identified with the copper mines in its immediate vicinity. There are several mines worked to a large extent, besides others of less note which will, no doubt, soon be rendered productive. The mineral range of Keweenaw Point, some 4 to 6 miles in width, extends through all this section of country, being as yet only partially explored. The Isle Royal, Huron, and Portage, are the principal mines worked on the south side of the lake.

PORTAGE LAKE is an irregular body of water about 20 miles in length, extending nearly across Keweenaw Point to within 2 miles of Lake Superior. Steamers and sail-vessels drawing 12 feet can pass through Portage Entry, and navigate the lake with safety. This body of water was an old and favorite thoroughfare for the Indians, and the Jesuit Fathers who first discovered and explored this section of country. A canal of two miles in length would render this portage route navigable for steamers and sail-vessels navigating Lake Superior, thereby reducing the distance over 100 miles. During the winter months the atmosphere is very clear and transparent in the vicinity of Houghton, and all through Keweenaw Point; objects can be seen at a great distance of a clear day, while sounds are conveyed distinctly through the atmosphere. presenting a phenomenon peculiar to all northern latitudes. This is the season of health and pleasure to the permanent residents.

HANCOCK, Houghton county, Michigan, is situated on the north side of Portage Lake, opposite to the village of Houghton, with which it is connected by a steam ferry. The town was first laid out in 1858, and now contains about 4,000 inhabitants, including the mining population on the north side of the lake; its sudden rise and prosperity being identified with the rich deposit of native copper, in which this section of country abounds.

The site of the village is on a side-hill rising from the lake level to a height of about 500 feet, where the opening to the mines is situated. Here is 1 Congregational, 1 Methodist, and 1 Roman Catholic church; 3 public-houses, the *Mason House* being a well-kept hotel; a number of stores and warehouses, 1 steam saw-mill, 1 barrel-factory, 1 foundry and machine-shop, and other manufacturing establishments; also, in the vicinity are 4 extensive steam stamping-mills worked by the different mining companies. The *Portage Lake Smelting Works* is an incorporated company, turning out annually a large amount of pure merchantable copper. The business of the company consists of fusing and converting the mineral into refined metal for manufacturing purposes.

The Quincy, Hancock, Pewabic, and Franklin mines are situated on the north side of the lake, on elevated ground. being now in active operation, giving employment to about 1,800 operatives. The successful working of these mines by means of improved machinery, in connection with the smelting works, will, no doubt, give profitable employment to thousands of miners and laborers, thereby rendering this locality the great copper mart of this region, the pure metal being shipped to the Eastern market during the season of navigation.

KEWEENAW POINT * is a large extent of land jutting out into Lake Superior, from 10 to 25 miles wide and about 60 miles in length. This section of country for upward of 100 miles, running from southwest to northeast, abounds in silver

* " On many maps spelled *Kewcewaiwona*, and otherwise. Pronounced by our Indians, ‘Ki-wi-wal-non-ing,’ now written and pronounced as above; meaning a portage, or place where a portage is made—the whole distance of some eighty or ninety miles around the Point being saved by entering Portage Lake and following up a small stream, leaving a portage of only about a half mile to Lake Superior on the other side."—*Foster and Whitney's Report.*

and copper ores, yielding immense quantities of the latter; much of it being pure native copper, but often in such large masses as to render it almost impossible to be separated for the purpose of transportation. Masses weighing from 1,000 to 5,000 pounds are often sent forward to the Eastern markets. The geological formation is very interesting, producing specimens of rare beauty and much value. MANITOU ISLAND lies off Keweenaw Point, on which is a light-house to guide the mariner to and from Copper Harbor. The island is about 7 miles in length and four wide.

COPPER HARBOR, Mich., is situated near the extreme end of Keweenaw Point, in N. lat. 47° 30′, W. long. 88° 00′; the harbor, although somewhat difficult to enter, is one of the best on Lake Superior, being distant 250 miles from the Saut Ste. Marie. The settlement contains about 200 inhabitants, a church, a hotel, and two or three stores. *Fort Wilkins*, formerly an U. S. military post, has been converted into a hotel, being handsomely situated on *Lake Funny Hoe*, about half a mile distant from the steamboat landing. In the vicinity are copper mines which have been extensively worked, and are well worthy of a visit.

AGATE HARBOR, 10 miles west of Copper Harbor, is the name of a small settlement. This port is not as yet much frequented by steamers.

EAGLE HARBOR, 16 miles west of Copper Harbor, is a good steamboat landing. Here are two churches, a good public-house, together with several stores and storehouses. Population about 700, being mostly engaged in mining. The Central, Copper, Falls, Pennsylvania, and Amygdaloid are the principal working copper mines.

EAGLE RIVER HARBOR and Village, eight miles further, are favorably situated at the mouth of a stream of the same name. Here are two churches, a well-kept hotel,

four stores and several storehouses. Population 800. This is a thriving settlement, it being the outport of the celebrated Cliff, or Pittsburgh and Boston, and other mines. The copper found in this vicinity is of the purest quality, where is found silver in small quantities, some of the specimens being highly prized. Off this harbor the lamented Dr. Houghton was drowned, October, 1845, while engaged in exploring this section of country: Keweenaw Point and adjacent country being very appropriately named Houghton County in honor of his memory.

On the north side of Keweenaw Point bold shores extend to near Ontonagon, with high lands in the distance, forming the rich copper range of this region.

Ontonagon, Ontonagon Co., Mich., 336 miles from the Saut Ste. Marie, is advantageously situated at the mouth of the river of the same name. The river is about 200 feet wide at its mouth, with a sufficient depth of water over the bar for large steamers. Here is being erected an extensive pier and breakwater. The village contains an Episcopal, a Presbyterian, and a Roman Catholic church; two good hotels, the *Bigelow House* and *Johnson House;* two steam saw-mills, and ten or twelve stores and storehouses, and about 1,200 inhabitants.

In this vicinity are located the Minnesota, the National, the Rockland, and several other very productive copper mines. The ore is found from twelve to fifteen miles from the landing, being imbedded in a range of high hills traversing Keweenaw Point from N. E. to S. W. for about 100 miles. Silver is here found in small quantities, beautifully intermixed with the copper ore, which abounds in great masses.

A good plank road runs from Ontonagon to near the Adventure Mine, and other mines, some twelve or fourteen miles distant, where commences the copper range of hills. A small steamer also runs on Ontonagon River to near the Minnesota and National Mines, where is a flourishing settlement inhabited by minors.

The *Ontonagon River* is thus beautifully described by ROBERT ALAN, Esq.,

To the Ontonagon River.

Sweet river, on thy silvery tide
The sable warriors no more glide;
Along thy wild and wooded shore
Their kindling watch-fires blaze no more.
Where'er thou roam'st by dale or hill
Thy banks are silent now and still,
As if thy waves, since time began,
Had ne'er been stained by savage man.
Unlike the tide of human time,
Which keeps each grief, retains each crime,
And deeper, as it downward flows,
Is stained with past and present woes.
Flow on, thou gentle river, flow
Through summer's rain and winter's snow ;
May Indian war-whoops no more wake
Thy echoes, as thou seek'st the lake,
But peaceful lovers by thy stream
On future joys and pleasures dream.
ST. ANDREW.

Population of Ontonagon County, 1860.

Towns, &c.	Males.	Females.	Total.
Algonquin,	46	31	77
Flint Steel,	20	10	30
Greenland,	105	67	296
Maple Grove,	67	57	
Minnesota,	660	184	844
National,	246	90	336
Nebraska,	34	22	56
Ontonagon,	650	498	1,148
Pewabic,	71	38	109
Rockland,	187	95	282
Rockland Mine,	206	47	253
Rosendale,	344	251	595
Superior,	15	9	24
Webster,	261	79	340
Williamsburg,	68	30	98

Total, 4,488

The PORCUPINE MOUNTAIN, lying some 15 or 20 miles west of Ontonagon, is a bold headland that can distinctly be seen at a great distance, rising some 1,300 feet above the lake surface.

Lake Superior Copper Mining Companies.

Name.	Agent.	President.	Office.
ADVENTURE,*	Thos. W. Buzzo,	C. G. Hussey,	Pittsburgh, Pa.
ALBANY & BOSTON,†	A. B. Wood,	Horatio Bigelow,	Boston, Mass.
AMYGDALOID,‡	A. C. Davis,	George L. Oliver,	Philadelphia.
AZTEC,*	Thos. W. Buzzo,	C. G. Hussey,	Pittsburgh, Pa.
BOHEMIAN,*	Wm. E. Dickenson,	W. R. Griffith,	New York.
CALEDONIA,*	Mr. Burgess,	T. F. Mason,	New York.
CARP LAKE,*		Fayette Brown,	Cleveland, Ohio.
CENTRAL,‡	C. B. Petrie,	J. L. Mott,	New York.
CLARKE,*|	Wm. Kirby,		Paris, France.
COPPER FALLS,‡	John Usen,	Horatio Bigelow,	Boston, Mass.
EAGLE RIVER,§		A. W. Spencer,	Boston, Mass.
EVERGREEN BLUFF,*	E. C. Roberts,	F. E. Eldred,	
FLINT STEEL RIVER,*	E. C. Roberts,	Charles E. Smith,	New York.
FRANKLIN,†	J. H. Foster,	Jerome Merritt,	Boston, Mass.
GARDEN CITY,§	G. W. Gatiss,	John M. Wilson	Chicago, Ill.

Name.	Agent.	President.	Office.
HANCOCK,†	Jonathan Cox,	A. Shurtleff,	New York.
HILTON,*	C. M. Sanderson,	T. F. Mason,	New York.
HURON,†	—— Collom,	Wm. Haywood,	Boston, Mass.
ISLE ROYALE,†	C. F. Eschweiler,	T. H. Perkins,	Boston, Mass.
INDIANA,		Wm. Harris,	—— ——
KNOWLTON,*	C. M. Sanderson,	W. J. Gordon,	Cleveland, Ohio.
MANHATTAN,	J. F. Blandy,	R. H. Rickard,	New York.
MANDAN,¶	A. B. Wood,	George L. Oliver,	Philadelphia.
MESNARD,†	Jacob Houghton, Jr.,	Horatio Bigelow,	Boston, Mass.
MICHIGAN,¶	A. B. Wood,	T. F. Mason,	New York.
MINNESOTA,*	J. B. Townsend,	Wm. Pearsall,	New York.
NATIONAL,*	Wm. Webb,		Pittsburgh, Pa.
NORWICH,*	E. C. Roberts,	A. H. Center,	New York.
OGIMA,*	Wm. W. Spalding,		—— ——
PETHERICK,‡	John Usen,	Horatio Bigelow,	Boston, Mass.
PENNSYLVANIA,‡	S. W. Hill,	Jos. G. Henszey,	Philadelphia.
PEWABIC,†	J. H. Foster,	William Haywood,	Boston, Mass.
PITTSBURGH & BOSTON (CLIFF),§	James Watson,	C. G. Hussey, ·	Pittsburgh, Pa.
PHŒNIX,§	O. A. Farwell,	John Jackson,	Boston, Mass.
PONTIAC,†	Jacob Houghton, Jr.,	Horatio Bigelow,	Boston, Mass.
PORTAGE,†	C. C. Douglass,	Thos. W. Lockwood,	Detroit, Mich.
QUINCY,†	S. S. Robinson,	Thomas F. Mason,	New York.
ROCKLAND,*	J. B. Townsend,	Samuel J. W. Barry,	New York.
STAR,¶	L. W. Clarke,		Boston, Mass. ·
SOUTH SIDE,†	C. F. Eschweiler,	Thomas H. Perkins,	Boston, Mass.
SUPERIOR,*	J. B. Townsend,	William Hickok,	New York.
TREMONT,*		Jerome Merritt,	Boston, Mass.
TOLTEL,*	Henry Buzzo,	L. W. Clarke,	Boston, Mass.
VICTORIA,*		Jerome Merritt,	Boston, Mass.

OUTPORTS.

* Ontonagon.   † Portage Lake.   ‡ Eagle Harbor.   § Eagle River.   ¶ Copper Harbor.

LA POINTE, 77 miles west of Ontonagon, situated on the south end of Madeline Island, the largest of the *Apostle Islands*, is one of the oldest settlements on Lake Superior; it was first peopled by the French Jesuits and traders in 1680, being 420 miles west of the Saut Ste. Marie, which was settled about the same time. The mainland and islands in this vicinity have been for many ages the favorite abode of the American Indian, now lingering and fading away as the country is being opened and settled by the white race.

The village now contains 300 inhabitants, most of whom are half-breeds and French. Here is an old Roman Catholic church, and one Methodist church; 2 hotels, 2 stores, and several coopering establishments for the making of fish-barrels.

The harbor and steamboat landing are on the south end of the island, where may usually be seen fishing-boats and other craft navigating this part of Lake Superior. Wheat, rye, barley, oats, peas, potatoes and other vegetables, are raised in large quantities. Apples, cherries, gooseberries and currents are raised in the gardens at La Pointe. The wild fruits are plums, cranberries, strawberries, red raspberries, and whortleberries. The principal forest-trees on the islands are maple, pine, hemlock, birch, poplar, and cedar trees.

BAYFIELD, capital of La Pointe Co., Wis., is favorably situated on the southern shore of Lake Superior, 80 miles east of its western terminus, and 3 miles west of La Pointe, being 80 miles west of Ontonagon. The harbor is secure and capacious, being protected by the Apostle Islands, lying to the northeast. The town plot rises from 60 to 80 feet above the waters of the lake, affording a splendid view of the bay, the adjacent islands and headlands. Its commercial advantages are surpassed by no other point on Lake Superior, being on the direct route to St. Paul, Minn., and the Upper Mississippi. Here are situated a Presbyterian, a Methodist, and a Roman Catholic church; 2 hotels, 4 stores, 2 warehouses, 1 steam saw-mill, and several mechanics' shops. Population in 1860, 300.

The *Hudson and Bayfield Railroad*, 164 miles in length, has been surveyed and will most probably be completed within a few years, there being a favorable land grant conceded to the company. This will afford a speedy route to St. Paul and other ports on the Mississippi River.

LA POINTE BAY, on the west side of which is situated the port of Bayfield, is a large and safe body of water, being protected from winds blowing from every point of the compass. The shores of the islands and mainland are bold, while the harbor affords good anchorage for the whole fleet of the lakes.

The Indian Agency for the Chippewa tribe of Indians residing on the borders of Lake Superior, have their headquarters at Bayfield. The annual annuities are usually paid in August of each year, when large numbers flock to the Agency to obtain their pay in money, provisions, and clothing.

ASHLAND, 12 miles south of La Pointe, at the head of Chagwamegon Bay, is another new settlement no doubt destined to rise to some importance, it having a very spacious and secure harbor.

MASKEG RIVER, a considerable stream, the outlet of several small lakes, enters Lake Superior about 15 miles east of Ashland; some 10 miles farther eastward enters MONTREAL RIVER, forming the boundary, in part, between the States of Michigan and Wisconsin.

The TWELVE APOSTLES' ISLES consist of the Madeline, Cap, Line, Sugar, Oak, Otter, Bear, Rock, Cat, Ironwood, Outer, and Presque Isle, besides a few smaller islands, being grouped together a short distance off the mainland, presenting during the summer months a most picuresque and lovely appearance. Here are to be seen clay and sandstone cliffs rising from 100 to 200 feet above the waters, while most of the islands are clothed with a rich foliage of forest-trees.

THE TWELVE APOSTLES' ISLANDS.

The following description of these romantic islands is copied from *Owen's Geological Survey of Wisconsin, &c.*

" When the waters of Lake Superior assumed their present level, these islands were doubtless a part of the promontory, which I have described as occupying the space between Chagwamegon Bay and Brute River. They are composed of drift-hills and red clay, resting on sandstone which is occasionally visible. In the lapse of ages, the winds, waves, and currents of the lakes cut away channels in these soft materials, and finally separated the lowest parts of the promontory into islands, and island-rocks, now twenty-three in number, which are true outliers of the drift and sandstone.

"At a distance they appear like mainland, with deep bays and points, gradually becoming more elevated to the westward. 'Ile au Chêne,' or Oak Island, which is next the Detour (or mainland), is a pile of detached drift, 250 or 300 feet high, and is the highest of the group. Madeline, 'Wau-ga-ba-me' Island, is the largest (on which lies La Pointe), being 13 miles long, from northeast to southwest, and has an average of 3 miles in breadth. "Muk-quaw" or Bear Island, and "Eshquagendeg" or Outer Islands, are about equal in size, being six miles long and two and a half wide.

"They embrace in all, an area of about 400 square miles, of which one-half is water. The soil is in some places good, but the major part would be difficult to clear and cultivate. The causes to which I have referred, as giving rise to thickets of evergreens along the coast of the lake, operate here on all sides, and have covered almost the whole surface with cedar, birch, aspen, hemlock, and pine. There are, however, patches of sugar-tree land, and natural meadows.

"The waters around the islands afford excellent white fish, trout, and siskowit, which do not appear to diminish after many years of extensive fishing for the lower lake markets. For trout and siskowit, which are caught with a line in deep water, the best ground of the neighborhood is off Bark Point or 'Point Ecorce' of the French. Speckled or brook trout are also taken in all the small streams.

"That portion of the soil of the islands fit for cultivation, produces potatoes and all manner of garden vegetables and roots in great luxuriance. In the flat wet parts, both the soil and climate are favorable to grass; and the crop is certain and stout. Wheat, oats, and barley do well on good soil when well cultivated.

"In regard to health, no portion of the continent surpasses the Apostle Islands. In the summer months they present to the residents of the South the most cool and delightful resort that can be imagined, and for invalids, especially such as are affected in the lungs or liver, the uniform bracing atmosphere of Lake Superior produces the most surprising and beneficial effects."

Healthy Influence of Lake Superior.

No better evidence can be given of the healthy climate of the Lake Superior region than the following extracts from letters, written by well-known individuals:

"BAYFIELD, July 28th, 1860.

"Dear Sir:—Perhaps it would be interesting to you to state, in a few words, the happy effects that this climate has produced for me.

"Previous to my coming here I consulted with three physicians in Philadelphia, one in the central part of Pennsylvania, one in Washington, D. C., and one in Georgetown, D. C. It was the opinion of all that consumption was tightening her grasp upon me, and that soon I would be laid in the grave. Under medical advice I made use of an inhaling apparatus, drank cod-liver oil and whiskey, but all without any beneficial results. Through the advice of friends, and in hopes of saving my life, I came to this place, June 6th, 1857, bringing with me three gallons of cod-liver oil and three gallons of old rye whiskey. This bracing atmosphere seemed to give me immediate relief, and in a short time it seemed as if a heavy load was removed from my chest. I used the cod-liver oil in feed for young chickens and greasing my boots, and gave the most of the whiskey away. I am now (three years after my arrival here) enjoying excellent health.

"Respectfully yours,

"J. H. N."

BRONCHIAL, OR THROAT DISEASE.

Rev. W—— L—— resided in Malone, Franklin county, New York, during the year 1850, where he first was troubled by the *bronchial disease*, which led to bleeding of the throat. From Malone he removed to Fairfield county, Connecticut, in 1852, near Long Island Sound, where the disease increased in virulence, assuming an alarming character. In 1855 he removed to Syracuse, New York, where he contracted a remittent fever, without being benefited in regard to his throat disease. In 1858 he visited Europe for the benefit of his health, without his throat disease being benefited, although he improved in general health. In August, 1859, he removed to Eagle River, Michigan, situated on the south shore of Lake Superior, where he gradually improved in health; but on moving a few miles in the interior, near one of the copper mines, his health rapidly improved, and a permanent cure was effected, as he supposes, by pure and bracing air—for which this whole section of country is justly celebrated.

Dated on board steamer NORTH STAR, July, 1860.

On proceeding from La Pointe westward, the steamer usually passes around Point de Tour, ten miles north, and enters Fond du Lac, a noble bay situated at the head of Lake Superior. It may be said to be 50 miles long and 20 miles wide, abounding in good fishing-grounds.

Superior, or SUPERIOR CITY, Douglass county, Wisconsin, is most advantageously situated on a bay of Superior, at the west end of the lake, near the mouth of St. Louis River. Here are a church, two hotels, and ten or fifteen stores and storehouses, and about 1,000 inhabitants. A small river, called the Nemadji, runs through Superior, and enters into St. Louis Bay. Perhaps no place on Lake Superior has commercial advantages equal to this town; its future is magnified almost beyond conception. The *St. Croix and Superior Railroad* is proposed to terminate at this place, extending southward to Hudson, on the St. Croix River, about 140 miles. Another railroad is proposed to extend westward to the Sa k Rapids, on the Upper Mississippi, either from this place or Portland, Minn.

DISTANCES FROM FOND DU LAC TO ST. PAUL, MINN.

Fond du Lac (St. Louis River)............		Miles.
Pokagema, (*Portage*)............... ...		75
Falls St. Croix (*Canoe*)............ .	40	115
Marine Mills, (*Steamboat*).............	19	134
Stillwater. "	11	145
St. Paul (*Stage*)........................	18	163

Distance from SUPERIOR CITY to ST. CLOUD (Sauk Rapids), by proposed railroad route, 120 miles. St. Cloud to ST. PAUL, 76 miles. Total, 196 miles.

DISTANCES FROM SUPERIOR CITY TO PEMBINA, MINN.

Superior........................		Miles.
Crow Wing		80
Otter Tail Lake...................	70	150
Rice River....................	74	224
Sand Hills River.................	70	294
Grand Fork (Red River).............	40	334
Pembina	80	414

From St. Paul to Pembina, *via* Crow Wing, 464 m.

FOND DU LAC, St. Louis county, Minn., is situated on St. Louis River, 20 miles above its entrance into Lake Superior. Vessels of a large class ascend to this place, being within four miles of the St. Louis Falls, having a descent of about 60 feet, affording an immense water-power. Here are sandstone and slate quarries, from which stone and slate are quarried, and extensively used for building purposes. Iron and copper ore abound in the vicinity. These advantages bid fair to make this point a mart of commerce and manufacture.

St. Louis River, flowing into the S. W. end of Lake Superior, is a large and important stream, and is navigable for steamers and lake craft for upward of 20 miles from its mouth. Above the falls (where the water has a descent of 60 feet, presenting a beautiful appearance), the river is navigable for canoes and small craft for about 80 miles farther. This river is the recipient of the waters of several small lakes lying almost due north of its outlet, its head waters flowing south from near Rainy Lake.

Portland, St. Louis county, Minn., advantageously situated at the extreme west end of Lake Superior, seven miles N. W. from Superior City, is a place of growing importance, where is a good steamboat landing, with bold shore. This is the capital of the county, and bids fair to be a successful competitor with Superior City for the carrying trade of the Great West and Pacific coast. Along the shore of the lake northward are to be seen bold sandy bluffs and highlands, supposed to be rich in mineral wealth.

Bellville, Minn., is a new settlement, situated on the lake shore, 4 or 5 miles north of Portland.

Clifton, St. Louis Co., Minn., situated 11 miles N. E. of the head of Lake Superior, is a new settlement. In the vicinity are rich copper mines and good farming lands.

Buchanan is another new settlement, situated northeast of Clifton, possessing similar advantages.

Burlington is a new settlement, situated near Agate Bay.

Encampment is the name of a river, island, and village, where is a good harbor, the mouth of the river being protected by the island. On the river, near its entrance into the lake, are falls affording fine water-power. Cliffs of greenstone are to be seen, rising from 200 to 300 feet above the water's edge, presenting a handsome appearance. To the north of Encampment, along the lake shore, abound porphyry and greenstone. This locality is noted for a great agitation of the magnetic needle; the depth of water in the vicinity is too great for vessels to anchor; the shores being remarkably bold, and in some places rising from 800 to 1,000 feet above the water.

Hiawatha is another new settlement, situated on the west shore of Lake Superior, where are found copper ore and other valuable minerals, precious stones, etc.

Beaver Bay, on the N. W. lake shore, at the mouth of Beaver River, affords a good harbor, where is a small settlement.

Grand Portage, Minn., advantageously situated on a secure bay, near the mouth of Pigeon River, is an old station of the American Fur Company. Here are a Roman Catholic Mission, a block-house, and some 12 or 15 dwellings. Mountains from 800 to 1,000 feet are here seen rising abruptly from the water's edge, presenting a bold and sublime appearance.

Pigeon Bay and River forms the northwest boundary between the United States and Canada, or the Hudson Bay Company's territory. Pigeon River is but a second-class stream, and by its junction with Arrow River continues the boundary through Rainy Lake and River to the Lake of the Woods, where the 49th degree of north latitude is reached. The mouth of Pigeon River is about 48 degrees north latitude, and 89 degrees 30 minutes west from Greenwich.

Along the whole west shore of Lake Superior, from St. Louis River to Pigeon River, are alternations of metamorphosed schists and sandstone, with volcanic grits and other imbedded traps and porphyry, with elevations rising from 800 to 1,200 feet above the lake, often presenting a grand appearance.

ISLE ROYALE, Houghton Co., Mich., being about 45 miles in length from N. E. to S. W., and from 8 to 12 miles in width, is a rich and important island, abounding in copper ore and other minerals, and also precious stones. The principal harbor and only settlement is on *Siskowit Bay*, being on the east shore of the island, about 50 miles distant from Eagle Harbor, on the main shore of Michigan.

The other harbors are—Washington Harbor on the southwest, Todd's Harbor on the west, and Rock Harbor and Chippewa Harbor on the northeast part of the island. In some places on the west are perpendicular cliffs of green-stone, very bold, rising from the water's edge, while on the eastern shore conglomerate rock or coarse sandstone abounds, with occasional stony beach. On this coast are many islets and rocks of sandstone, rendering navigation somewhat dangerous. Good fishing-grounds abound all around this island, which will, no doubt, before many years, become a favorite summer resort for the invalid and sportsman, as well as the scientific tourist.

SISKOWIT LAKE is a considerable body of water lying near the centre of the island, which apparently has no outlet. Other small lakes and picturesque inlets and bays abound in all parts of the island. Hills, rising from 300 to 400 feet above the waters of the lake, exist in many localities throughout the island, which is indented by bays and inlets.

Northern Shore of Lake Superior.

EXTRACT *from Report on the Geology of the Lake Superior Country, by* FOSTER *and* WHITNEY:

NORTHERN SHORE. — "Beginning at Pigeon Bay, the boundary between the

9

United States and the British Possessions (north latitude 48°), we find the eastern portion of the peninsula abounds with bold rocky cliffs, consisting of trap and red granite.

"The Falls of Pigeon River, eighty or ninety feet in height, are occasioned by a trap dyke which cuts through a series of slate rocks highly indurated, and very similar in mineralogical characters to the old graywacke group. Trap dykes and interlaminated masses of traps were observed in the slate near the falls.

"The base of nearly all the ridges and cliffs between Pigeon River and Fort William (situated at the mouth of Kaministequoi River, the western boundary of Upper Canada) is made up of these slates, and the overlaying trap. Some of the low islands exhibit only the gray grits and slates. Welcome Islands, in Thunder Bay, display no traps, although, in the distance, they resemble igneous products, the joints being more obvious than the planes of stratification, thus giving a rude semi-columnar aspect to the cliffs.

"At Prince's Bay, and also along the chain of Islands which lines the coast, including Spar, Victoria, and Pie islands, the slates with the crowning traps are admirably displayed. At the British and North American Company's works the slates are traversed by a heavy vein of calc-spar and amethystine quartz, yielding gray sulphuret and pyritous copper and galena. From the vein where it cuts the overlaying trap on the main shore, considerable silver has been extracted.

"At Thunder Cape, the slates form one of the most picturesque headlands on the whole coast of Lake Superior. They are made up of variously colored beds, such as compose the upper group of Mr. Logan, and repose in a nearly horizontal position. These detrital rocks attain a thickness of nearly a thousand feet, and are crowned with a sheet of trappean rocks three hundred feet in thickness.

"At L'Anse à la Bouteille (opposite the Slate Islands, on the north shore of Lake Superior) the slates reappear, with the granite protruding through them, and occupy the coast for fifteen miles; numerous dykes of greenstone, bearing east and west, are seen cutting the rocks vertically. The Slate Islands form a part of this group, and derive their name from their geological structure.

"They are next seen, according to Mr. Logan, for about seven miles on each side of the Old Pick River. Near Otterhead a gneissoidal rock forms the coast, which presents a remarkably regular set of strata in which the constituents of sienite are arranged in thin sheets and in a highly crystalline condition. From this point to the Michipicoten River the slates and granite occupy alternate reaches, along the coast, for the distance of fifty miles. ' With the exception of a few square miles of the upper trap of gargantua, these two rocks appear to hold the coast all the way to the vicinity of Pointe aux Mines, at the extremity of which they separate from the shore, maintaining a nearly straight southeasterly line across the Batchewanung Bay, leaving the trap of Mamainse between them and the lake. Thence they reach the northern part of Goulais Bay, and finally attain the promontory of Gros Cap, where they constitute a moderately bold range of hills, running eastwardly toward Lake Huron.' "*

Fisheries of Lake Superior.

Good fishing-grounds occur all along the north shore of Lake Superior, affording a bountiful supply of white-fish, Mackinac trout, and many other species of the finny tribe. On the south shore there are fisheries at White-Fish Point, Grand

* Canadian Report, 1846-'47.

Island, near the Pictured Rocks, Keweenaw Point, La Pointe, and Apostles' Islands, and at different stations on Isle Royale, where large quantities are taken and exported; but there are no reliable statistics as to the number of men employed or the number of barrels exported. Between the head of Keweenaw Point and the mouth of the Outonagon River, considerable quantities of fish are taken, for which there is a ready market at the mining stations. In addition to the white fish and Mackinac trout, the siskowit is occasionally taken. Its favorite resort, however, is the deep water in the vicinity of Isle Royale.

LAKE SUPERIOR TROUT-FISHING IN WINTER.—The Lake Superior *Journal* says:

"Angling through the ice to a depth of thirty fathoms of water is a novel mode of fishing somewhat peculiar to this peculiar region of the world. It is carrying the war into fishdom with a vengeance, and is denounced, no doubt, in the communities on the bottom of these northern lakes as a scaly piece of warfare. The large and splendid salmon-trout of these waters have no peace; in the summer they are enticed into the deceitful meshes of the gill-net, and in the winter, when they hide themselves in the deep caverns of the lakes, with fifty fathoms of water above their heads, and a defence of ice two or three feet in thickness on the top of that, they are tempted to destruction by the fatal hook.

"Large numbers of these trout are caught every winter in this way on Lake Superior; the Indian, always skilled in the fishing business, knows exactly where to find them and how to kill them. The whites make excursions out on the lake in pleasant weather to enjoy this sport. There is a favorite resort for both fish and fishermen near Gros Cap, at the entrance of Lake Superior, through the rocky gateway between Gros Cap and Point Iroquois, about 18 miles above the Saut, and many a large trout, at this point, is pulled up from its warm bed at the bottom of the lake, in winter, and made to bite the cold ice in this upper world. To see one of these fine fish, four or five feet in length, and weighing half as much as a man, floundering on the snow and ice, weltering and freezing to death in its own blood, oftentimes moves the heart of the fisherman to expressions of pity.

"The *modus operandi* in this kind of great trout-fishing is novel in the extreme, and could a stranger to the business overlook at a distance a party engaged in the sport, he would certainly think they were mad, or each one making foot-races against time. A hole is made through the ice, smooth and round, and the fisherman drops down his large hook, baited with a small herring, pork, or other meat, and when he ascertains the right depth, he waits—with fisherman's luck—some time for a bite, which in this case is a pull all together, for the fisherman throws the line over his shoulder, and walks from the hole at the top of his speed till the fish bounds out on the ice. We have known of as many as fifty of these splendid trout caught in this way by a single fisherman in a single day: it is thus a great source of pleasure and a valuable resource of food, especially in Lent, and the most scrupulous anti-pork believers might here 'down pork and up fish' without any offence to conscience."

List of Vessels Lost in the Lake Superior Trade.

Since the discovery of copper in the Upper Peninsula, in 1845, and the commencement of the Lake Superior steamer and vessel trade, many craft engaged in the trade have been lost. Previous to the discovery of copper, there was no other trade but that in furs, and one of the fur-trading vessels was lost—the John Jacob Astor. We have compiled the following table, which will be found of interest to those connected with the Lake Superior copper trade:—

Name of Vessel lost.	Value.	Value Cargo.	Year
Schooner Merchant...	$4,000	$2,000	1847
Propeller Goliath...	18,000	18,000	1847
Steamer Ben Franklin...	15,000	4,000	1850
Propeller Monticello...	30,000	10,000	1851
Schooner Siskowit...	1,000	500	....
Propeller Independence.	12,000	15,000	1853
Steamer Albany...	30,000	2,500	1853
Propeller Peninsula...	18,000	12,000	1854
Steamer E. K. Collins...	100,000	1,500	1854
Steamer Baltimore...	15,000	4,000	1855
Steamer Superior...	15,000	10,000	1856
Propeller B. L. Webb...	50,000	15,000	1856
Propeller City of Superior.	50,000	25,000	1857
Propeller Indiana...	6,000	2,500	1858
	$366,000	$125,000	

—making a grand total of $491,000.

Since the above Table was compiled the following losses have occurred in the Lake Superior trade:

Steamer Arctic, wrecked on Lake Superior, June, 1860.

Steamer Gazelle, wrecked on Lake Superior, 1860.

Steamer Elgin, lost on Lake Michigan, September 7, 1860.

Steamer North Star, burnt at Cleveland, February, 1862.

The loss of life by the accidents given above is, as near as can be ascertained, as follows:—

Schooner Merchant	18
Propeller Independence	3
Steamer E. K. Collins	20
Steamer Superior	54
Steamer Lady Elgin	350
Total	445

There have been numerous losses of freight by jetisons and otherwise, that are not included in the table we have given,—and, what is rather singular, almost the whole of the jettisons and losses of hulls and cargoes have occurred while the vessels have been upward bound. *Detroit Advertiser.*

The Lakes--Land of the Free.

Columbia's shores are wild and wide,
  Columbia's *Lakes* are grand,
And rudely planted side by side,
  Her forests meet the eye;
But narrow must those shores be made,
  And low Columbia's hills,
And low her ancient forests laid,
  Ere *freedom* leaves her fields;
For 'tis the land where, rude and wild,
She played her gambols when a child.

And deep and wide her streams that flow
  Impetuous to the tide,
And thick and green the laurels grow
  On every river side;
But should a trans-Atlantic host
  Pollute our waters fair,
We'll meet them on the rocky coast,
  And gather laurels there;
For O, Columbia's sons are brave,
And free as ocean's wildest wave.

The gale that waves her mountain pine
  Is fragrant and serene,
And never brighter sun did shine
  Than lights her valleys green;
But putrid must those breezes blow,
  The sun must set in gore,
Ere footsteps of a foreign foe
  Imprint Columbia's shore;
For O, her sons are brave and free,
Their breasts beat high with Liberty.

The Land of Lake and River.

Composed by Dr. Laycock, of Woodstock,
  C. W.—A CANADIAN SONG.

The Land of Lake, River, and Forest wide,
Where Niagara plunges in splendor and pride
O'er the trembling cliffs her precipitous tide;
    Know ye the land?
      'Tis a glorious land!
And the land is our own dear home

The land which nor Arts nor Industry graced,
Where the bountiful seasons ran all to waste,
Till the Briton the Savage and Sluggard displaced;
    Know ye the land, &c.

The land where the Saxon, the Gaul, and the Celt.
The first glow of patriot brotherhood felt,
And forgetting old feuds in amity dwelt;
    Know ye the land, &c.

The land unpolluted by Despot or Slave,
Where justice is done on the Dastard and Knave,
Where honor is paid to the Wise and the Brave:
    Know ye the land, &c.

The land where the *Teacher* is honored and sought;
Where the *Schools* are all busy, the children all taught;
Where the *Thinker*, unfettered, can utter his thought;
    Know ye the land, &c.

The land where the *Farmer* is Lord of the Soil,
Where the *Toiler* himself reaps the fruit of his toil,
Where none has a *Title* his neighbor to spoil;
    Know ye the land, &c.

The land where the *Christian* can openly pray,
As Scripture and Conscience may show him the way,
Fearless of clerical tyrant or lay;
    Know ye the land, &c.

The land which, the older and stronger it grew,
To Law and to Loyalty still kept more true,
Both to *Prince* and to *People* according their due;
    Know ye the land?
      'Tis a glorious land!
And the land is our own dear home!

Trip along the North Shore of Lake Superior,

MADE ON BOARD THE CANADIAN STEAMER PLOUGHBOY, AUGUST. 1860.

On leaving the mouth of the *Ship Canal*, above the Rapids at the Saut Ste. Marie, a beautiful stretch of the river is passed and *Waiska Bay* entered, which is a small expanse of water extending westward to *Point Iroquois*, on the south shore, 15 miles distant. Immediately opposite rises GROS CAP, on the Canada side, being about four miles asunder. This bold headland consists of hills of porphyry rising from 600 to 700 feet above the waters of the lake. "Gros Cap is a name given by the *voyageurs* to almost innumerable projecting headlands; but in this case appropriate—since it is the conspicuous feature at the entrance of the lake."

North of Gros Cap lies GOULAIS BAY, and GOULAIS POINT, another bold highland which is seen in the distance. *Goulais River* enters the bay, affording, in connection with the adjacent waters, good fishing-grounds; the brook or speckled trout being mostly taken in the river. Here is a large Indian settlement of the Chippewa tribe. The whole north shore, as seen from the deck of the steamer, presents a bold and grand appearance, while in the distance, westward, may be seen the broad waters of Lake Superior.

TAQUAMENON BAY is next entered, which is about 25 miles long and as many broad, terminating at *White-Fish Point*, 40 miles above Saut Ste. Marie. PARISIEN ISLAND is passed 30 miles from the Saut, lying near the middle of the above bay, being attached to Canada.

SANDY ISLANDS, lying off *Batcheewauung Bay*, form, with others, a handsome group of islands, where are good fishing-grounds, being distant from the Saut Ste. Marie about 35 miles.

MAMAINSE POINT (*Little Sturgeon*), opposite White-Fish Point, is another bold headland, where is a fishing station and a few dwellings. The Montreal Company's copper mine is located near this point, 45 miles north of the Saut, where is a small settlement of miners. Here is a good harbor, the land rising abruptly to the height of 300 feet, presenting a rugged appearance. Some 12 or 15 miles north are located, on MICA BAY, the Quebec Copper Mining Company's Works, at present abandoned, owing to their being found unproductive. Still farther north, skirting Lake Superior, is to be found a vast *Mineral Region*, as yet only partially explored.

MONTREAL ISLAND, and RIVER, 20 miles north of Mamainse, afford good fishing-grounds. Here is a harbor exposed to the west winds from off the lake, which can safely be approached when the winds are not boisterous.

LIZARD ISLAND and LEACH ISLAND, some 10 miles farther northward, are next passed, lying contiguous to the mainland.

CAPE GARGANTUA, 40 miles north of Mamainse, is a bold headland. On the south side is a harbor protected by a small island. From this cape to the island of Michipicoten the distance is about 30 miles.

MICHIPICOTEN HARBOR, and RIVER, 110 miles north of the Saut Ste. Marie, situated in N. lat. 47° 56', W. long. 85° 06', affords a safe anchorage, being surrounded by high hills. Here is established a Roman Catholic mission, and an important Hudson Bay Company's post, from whence diverges the river and portage route to James's Bay, some 350 miles distant. The shore of the Lake here tends westward toward *Otter Head*, about 50 miles distant, presenting a bold and rugged appearance. This post, no doubt, is destined to become a place of resort as well as a commercial depot, from whence is now distributed the merchandise belonging to the above gigantic company—having exclusive sway over an immense region of country, extending northward to the arctic regions, and westward to the Pacific Ocean.

MICHIPICOTEN ISLAND (the *Island of Knobs or Hills*), 65 miles from Mamainse Point in a direct course, running in a northwest direction, lies about 40 miles west of Michipicoten Harbor. This island, 15 miles in length and 6 miles wide, may be called the *gem* of Lake Superior, presenting a most beautiful appearance as approached from the southward, where a few picturesque islands may be seen near the entrance to a safe and commodious harbor, which can be entered during all winds. Nature seems to have adapted this island as a place of resort for the seekers of health and pleasure. Within the bay or harbor a beautiful cluster of islands adorns its entrance, where may be found agates and other precious stones; while inland is a most charming body of water, surrounded by wooded hills rising from 300 to 500 feet above the waters of Lake Superior. The shores of the island abound with greenstone and amygdaloid, while copper and silver mines are said to exist in the interior, of great value, although, as yet, but partially explored. The fisheries here are also valuable, affording profitable employment to the hardy fisherman of this region. As yet, but one single shanty is erected on the shores of this romantic island, where, sooner or later, will flock the wealthy and beautiful in search of health and recreation, such as are afforded by pure air, boating, fishing, and hunting.

The fish mostly taken in this part of the lake are white-fish, siskowit, Mackinac trout, and speckled trout, the former being taken by gill-nets.

On the mainland are found the carabou, a large species of deer, bears, foxes, otters, beavers, martins, rabbits, partridges, pigeons, and other wild game. The barberry, red raspberry, and whortleberry are also found in different localities.

CARIBOU ISLAND, lying about 25 miles south of Michipicoten, near the middle of the lake, is a small body of land attached to Canada. It is usually passed in sight when the steamers are on their route to Fort William.

OTTER BAY, 25 miles north of Michipicoten, is a beautiful and secure body of water, being protected by an island at its entrance. Here is a wild and rugged section of country, abounding in game of the fur-bearing species.

Other bays and islands are found along the north shore beyond Otter Head, toward Pic River and Island, and said to be of great beauty, the whole coast being bold and rugged as seen from the water. At the mouth of the Pic is situated a Hudson Bay Company's Post.

SLATE ISLANDS are a cluster of great interest, where is to be found a large and secure harbor, lying north of the principal island of the group. To the north, on the mainland, are numerous bays and inlets affording safe harbor. As yet, the wild savage of the north alone inhabits this section of Canada West, which no doubt is rich in minerals of different kinds. The Hudson Bay Company's vessels now afford the only means of visiting this interesting region, which can alone be brought into notice and settled by the discovery of copper or silver mines of value sufficient to induce capitalists to organize Mining Companies.

COPPER REGION OF LAKE SUPERIOR—
NORTH SHORE.

See *Whitney's Metallic Wealth of the United States*, Phila., 1854.

The North Shore of Lake Superior is supposed to be very rich in mineral productions, although as yet but partially explored. The "Montreal Mining Company" have a mine which is now being worked to a limited extent at Mamainse Point, affording gray sulphuret of copper of a rich quality. The "Quebec and Lake Superior Mining Association" commenced operations in 1846 at Mica Bay, a few miles north of Mamainse, on a vein said to be rich in gray sulphuret of copper. An adit was driven 200 feet, three shafts sunk, and the 10-fathom level commenced. After spending $30,000 it was discovered that the mines were unproductive, and the works were abandoned.

A number of localities were explored, and worked to some extent on Michipicoten Island and on the mainland to the northward, but they are now nearly all abandoned. A surveying party, however, are now (1860) engaged in exploring the north shore of Lake Superior, under the authority of the Provincial Parliament, in order to be able to report in regard to the mineral region.

The northwest borders of the lake, and in particular the Island of St. Ignace, Black Bay, Thunder Cape, Pie Island, and the vicinity of Prince's Bay are supposed to be rich in both copper and silver. Splendid crystallizations of amethystine quartz and calc spar have been obtained on Spar Island, near Prince's Bay, and at other localities.

FORT WILLIAM, an important Hudson Bay Company's Post, is advantageously situated at the mouth of the Kaministiquia River, in north latitude 48 degrees 23 minutes, west longitude 89 degrees 27 minutes. Here is a convenient wharf and safe harbor, the bar off the mouth of the river affording 7 or 8 feet of water, which can easily be increased by dredging. The Company's buildings consist of a spacious dwelling-house, a store, and 3 storehouses, besides some 10 or 12 houses for the accommodation of the attachés and servants in the employ of the above gigantic company. The land is cleared for a considerable distance on both sides of the river, presenting a thrifty and fertile appearance. Wheat, rye, oats, barley, potatoes, and most kinds of vegetables are here raised in abundance; also, grass and clover of different kinds. The early frosts are the great hindrance to this whole section of country, which is rich in minerals, timber, furs, and fish; altogether producing a great source of wealth to the above company. Pine, spruce, hemlock, cypress, and balsam trees are common, also white birch, sugar-maple, elm, and ash, together with some hardy fruit-bearing trees and shrubs.

The *Roman Catholic Mission*, situated 2 miles above the company's post, on the opposite side of the river, is an interesting locality. Here is a Roman Catholic church and some 50 or 60 houses, being mostly inhabited by half-breeds and civilized Indians, numbering about 300 souls. The good influence of the Roman Catholic priests, along the shores of Lake Superior are generally admitted by all unprejudiced visitors—the poor and often degraded Indian being instructed in agriculture and industrial pursuits, tending to elevate the human species in every clime.

McKay's Mountain, lying 3 miles west of Fort William, near the Roman Catholic Mission, presents an abrupt and grand appearance from the water, being elevated 1,000 feet. Far inland are seen other high ranges of hills and mountains, presenting altogether, in connection with the islands, a most interesting and sublime view.

KAMINISTIQUIA, or "*Gah-mahnatekwai-ahk*" River, signifying in the Chippewa language the, "*place where there are many currents*," empties its waters into Thunder Bay. This beautiful stream affords navigation for about 12 miles, when rapids are encountered by the ascending voyageur. Some 30 miles above its mouth is a fall of about 200 feet perpendicular descent.

THUNDER BAY presents a large expanse of water, being about 25 miles in length and from 10 to 15 miles wide, into which flows several small streams, abounding in speckled trout. *Thunder Cape*, on the east, is a most remarkable and bold highland, being elevated 1,350 feet above Lake Superior. It rises in some places almost perpendicular, presenting a basaltic appearance, having on its summit an extinct volcano. From the elevated portions of this cape a grand and imposing view is obtained of surrounding mountains, headlands, and islands—overlooking *Isle Royale* to the south, and the north shore from McKay's Mountain to the mouth of Pigeon River; near Grand Portage, Minnesota.

PIE ISLAND, in the Indian dialect called *"Mahkeneeng"* or *Tortoise*, bounding Thunder Bay on the south, is about 8 miles long and 5 miles wide, and presents a most singular appearance, being elevated at one point 850 feet above the lake. This bold eminence is shaped like an enormous *slouched hat*, or inverted pie, giving name to the island by the French or English explorers, while the Indians gave it the name of tortoise from its singular shape. This elevated point is basaltic, rising perpendicular near the top, like the *Palisades* of the Hudson River.

Thunder Bay, and its vicinity, has long been the favorite residence of Indian tribes who now roam over this vast section of country, from Lake Superior to Hudson Bay on the north. The mountain peaks they look upon with awe and veneration, often ascribing some fabulous legend to prominent localities. A learned Missionary, in describing this interesting portion of Lake Superior and its inhabitants, remarked, that "the old Indians were of the opinion that *thunder clouds* are large gigantic birds, having their nests on high hills or mountains, and who made themselves heard and seen very far off. The head they described as resembling that of a huge eagle, having on one side a wing and one paw, on the other side an arm and one foot. The lightning is supposed to issue from the extremity of the beak through the paw, with which they launch it forth in fiery darts over the surrounding country."

Black Bay, lying east of Thunder Cape, is 45 miles long, and from six to eight miles wide, being encompassed on both sides by high and lofty hills. Towards the north are two peaked eminences termed the *Mamelons* or *Pups*, from their singular formation, resembling a female's breast, when seen at a distance.

Neepigon Bay is another romantic sheet of water, containing a number of beautiful wooded islands. It is about forty miles long and fifteen miles wide, being mostly surrounded by high and rocky eminences. Here the explorer, hunter, and angler may alike enjoy themselves with the wonders of the surrounding scene. Copper, silver, lead, and precious stones are here to be found in abundance on the islands and the mainland; while there is no end to the game and fish of this region.

The *Island of St. Ignace*, lying on the north shore of Lake Superior, is a large and important body of land, being rich in minerals and precious stones. It is about 17 miles long and six miles wide; the hills rising to 1,300 feet in some places, giving it a wild and romantic appearance from the water. Here are five small lakes, all being connected, and the outlet forming a beautiful stream, with rapids and falls. Large quantities of brook trout make these lakes and streams their favorite resort, they being but frequently visited except by the trapper and miner. Copper, silver, and lead are said to be found on this island in large quantities, as well as on other islands in its vicinity. The whole archipelago and mainland here afford good and safe harbors; the Canada side of the lakes being greatly favored in this respect.

PORTAGE ROUTE FROM LAKE SUPERIOR TO LAKE WINNIPEG,

STARTING FROM FORT WILLIAM, C. W.

KAMINISTAQUOIAH RIVER, emptying into Thunder Bay of Lake Superior, forms the west boundary of Canada proper; to the north and west lies the extensive region or country known as the *Hudson Bay Company's Territory*. Here commences the great *Portage Road* to Rainy Lake, Lake of the Woods, and the Red River settlement; also, to Lake Winnipeg, Norway House, and York Factory, situated on Hudson Bay. At the mouth of the Kaministaquoiah stands *Fort William*. "The banks of the river average in height from eight to twenty feet; the soil is alluvial and very rich. The vegetation all along its banks is remarkably thrifty and luxuriant in its appearance. The land is well timbered; there are found in great abundance, the fir-tree, birch, tamarack, poplar, elm, and the spruce, There is also white pine, but not in great plenty. Wild hops and peas are found in abundance, and some bushes and other flowering shrubs, in many places cover the banks down to the very margin of the river, adorning them with beauty, and often filling the air with fragrance. The land on this river up to the Mountain Portage (32 miles), and for a long way back, is unsurpassed in richness and beauty by any lands in British America."

The *Mountain Fall*, situated on this stream, is thus described: "We had great difficulty in finding it at first, but, guided by its thundering roar, through such a thicket of brush, thorns and briars, as I never before thought of, we reached the spot from whence it was visible. The whole river plunged in one broad white sheet, through a space not more than fifty feet wide, and over a precipice higher, by many feet, than the *Niagara* Falls. The concave sheet comes together about three-fourths of the way to the bottom, from whence the spray springs high into the air, bedewing and whitening the precipitous and wild looking crags with which the fall is composed, and clothing with drapery of foam the gloomy pines, that hang about the clefts and fissures of the rocks. The falls and the whole surrounding scenery, for sublimity, wildness, and novel grandeur, exceeds any thing of the kind I ever saw."—*Rev. J. Ryerson's Tour.*

The danger of navigating these mountain streams, in a birch canoe, is greater than many would expect who had never witnessed the force of the current sometimes encountered. Mr. Ryerson remarks: "During the day we passed a large number of strong and some dangerous rapids. Several times the canoe, in spite of the most strenuous exertions of the men, was driven back, such was the violence of the currents. On one occasion such was the force of the stream, that though four strong men were holding the rope, it was wrenched out of their hands in an instant, and we were hurled down the rapids with violent speed, at the mercy of the foaming waves and irresistible torrent, until fortunately in safety we reached an eddy below." (*See Engraving.*)

DOG LAKE is an expansion of the river, distant by its winding course, 76 miles from its mouth. Other lakes and expansions of streams are passed on the route westward.

"The SAVAN, or PRAIRIE PORTAGE, 120 miles from Fort William, by portage route, forms the height of land between Lake Superior and the waters falling into Lake Winnipeg; it is between three and four miles long, and a continuous cedar swamp from one end to the other, and is therefore very properly named the *Savan* or *Swamp*

Portage. It lies seven or eight hundred feet above Lakes Superior and Winnipeg, and 1,483 feet above the sea."

The SAVAN RIVER, which is first formed by the waters of the Swamp, enters into the *Lac Du Mille*, or the Lake of Thousands, so called because of the innumerable islands which are in it. This lake is comparatively narrow, being sixty or seventy miles in length.

The *River Du Mille*, the outlet of the Lake, is a precipitous stream, whereon are several portages, before entering into Lac La Pluie, distant 350 miles from Fort William.

RAINY LAKE, or *Lac la Pluie*, through which runs the boundary between the United States and Canada, is a most beautiful sheet of water; it is forty-eight miles long, and averages about ten miles in breadth. It receives the waters flowing westward from the dividing ridge separating the waters flowing into Lake Superior.

RAINY LAKE RIVER, the outlet of the lake of the same name, is a magnificent stream of water; it has a rapid current and averages about a quarter of a mile in width; its banks are covered with the richest foliage of every hue; the trees in the vicinity are large and varied, consisting of ash, cedar, poplar, oak, birch, and red and white pines; also an abundance of flowers of gaudy and variegated colors. The climate is also very fine, with a rich soil, and well calculated to sustain a dense population as any part of Canada.

The LAKE OF THE WOODS, or *Lac Du Bois*, 68 miles in length, and from fifteen to twenty-five miles wide, is a splendid sheet of water, dotted all over with hundreds of beautiful islands, many of which are covered with a heavy and luxuriant foliage. Warm and frequent showers occur here in May and June bringing forth vegetation at a rapid rate, although situated on the 49th degree of north latitude, from whence extends *westward* to the Pacific

PULLING A CANOE UP THE RAPIDS.

Ocean, the boundary line between the United States and Canada.

"There is nothing, I think, better calculated to awaken the more solemn feelings of our nature, than these noble lakes studded with innumerable islets, suddenly bursting on the traveller's view as he emerges from the sombre forest rivers of the American wilderness. The clear, unruffled water, stretching out on the horizon; here intersecting the heavy and luxuriant foliage of an hundred woody isles, or reflecting the wood-clad mountains on its margin, clothed in all the variegated hues of autumn; and there glittering with dazzling brilliancy in the bright rays of the evening sun, or rippling among the reeds and rushes of some shallow bay, where hundreds of wild fowl chatter as they feed with varied cry, rendering more apparent, rather than disturbing the solemn stillness of the scene: all tend to raise the soul from nature up to nature's God, and remind one of the beautiful passage of Scripture, 'O Lord, how marvellous are thy works, in wisdom hast thou made them all; the earth is full of thy riches.'"—*Ballantyne*.

The WINNIPEG RIVER, the outlet of the Lake of the Woods, is a rapid stream, of large size, falling into Winnipeg Lake, 3 miles below *Fort Alexander*, one of the Hudson Bay Company's Posts. A great number of Indians resort to the Fort every year, besides a number of families who are residents in the vicinity, here being one of their favorite haunts.

Rev. Mr. Ryerson remarks :—"The scenery for many miles around is strikingly beautiful. The climate for Hudson's Bay Territory is here remarkably fine and salubrious, the land amazingly rich and productive. The water in Lakes Lac La Pluie, Lac Du Bois, Winnipeg, &c., is not deep, and because of their wide surface and great shallowness, during the summer season, they become exceedingly warm ; this has a wonderful effect on the temperature of the atmosphere in the adjacent neighborhoods, and no doubt makes the great difference in the climate (or at least is one of the principal causes of it), in those parts, to the climate and vegetable productions in the neighborhood of Lake Superior, near Fort William. They grow spring wheat here to perfection, and vegetation is rapid, luxuriant, and comes to maturity before frosts occur."

The whole region of country surrounding Lake Winnipeg, the Red River country, as well as the Assiniboine and Saskatchewan country, are all sooner or later destined to sustain a vigorous and dense population.

LAKE WINNIPEG,

Situated between 50° and 55° north latitude, is about 300 miles long, and in several parts more than 50 miles broad ; having an estimated area of 8,500 square miles.* Lake Winnipeg receives the waters of numerous rivers, which, in the aggregate, drain an area of 400,000 square miles. The *Saskatchewan* (the river that runs fast) is its most important tributary. The Assiniboine, the Red River of the North, and Winnipeg River are its other largest tributaries, altogether discharging an immense amount of water into this great inland lake. It is elevated about 700 feet above Hudson Bay, and discharges its surplus waters through *Nelson River*, a large and magnificent stream, which like the St. Lawrence is filled with islands and numerous rapids,

* LAKE BAIKAL, the most extensive body of fresh water on the Eastern Continent, situated in Southern Siberia, between lat. 51° and 55° north, is about 370 miles in length, 45 miles average width, and about 900 miles in circuit ; being somewhat larger than Lake Winnipeg in area. Its depth in some places is very great, being in part surrounded by high mountains. The *Yenisei*, its outlet, flows north into the Arctic Ocean.

preventing navigation entirely below Cross Lake.

Lakes Manitobah and *Winnipego-sis*, united, are nearly of the same length as Winnipeg, lying 40 or 50 miles westward. Nearly the whole country between Lake Winnipeg and its western rivals is occupied by smaller lakes, so that between the valley of the Assiniboine and the eastern shore of Winnipeg fully one-third is under water. These lakes, both large and small, are shallow, and in the same water area show much uniformity in depth and coast line.

Lakes in the Valley of the Saskatchewan.

	Length in miles.	Breadth in miles.	Elevation in feet.	Area in m's.
Winnipeg,	280	57	628	8,500
Manitobah,	122	24	670	2,000
Winnipego-sis,	120	27	692	2,000
St. Martin,	30	16	655	350
Cedar,	30	25	688	350
Dauphin,	21	12	700	200

All the smaller lakes lie west of Lake Winnipeg, which receives their surplus waters; the whole volume, with the large streams, flowing into *Nelson River*, discharges into Hudson Bay, near York Factory, in 57° north latitude. The navigation of the latter stream is interrupted by falls and rapids, having a descent of 628 feet in its course of about 350 miles.

"The climate in the region of the above lakes and the Red River Settlement will compare not unfavorably with that of Kingston and Toronto, Canada West. The Spring generally opens somewhat earlier, but owing to the proximity of Lake Winnipeg which is late of breaking up, the weather is always variable until the middle of May. The slightest breeze from the north or northwest, blowing over the frozen surface of that inland sea, has an immediate effect on the temperature during the Spring months. On the other hand, the Fall is generally open, with mild, dry, and pleasant weather."

Red River of the North.

This interesting section of country being closely connected with the Upper Lakes, and attracting much attention at the present time, we subjoin the following extract from "MINNESOTA AND DACOTA," by C. C. Andrews:

"It is common to say that settlements have not been extended beyond Crow Wing, Minnesota. This is only technically true. A few facts in regard to the people who live four or five hundred miles to the north will best illustrate the nature of the climate and its adaptedness to agriculture.

"There is a settlement at *Pembina*, near the 49th parallel of latitude, where the dividing line between British America and the United States crosses the Red River of the North. Pembina is said to have about 600 inhabitants. It is situated on the Pembina River. It is an Indian-French word meaning '*Cranberry*.' Men live there who were born there, and it is in fact an old settlement. It was founded by British subjects, who thought they had located on British soil. The greater part of its inhabitants are half-breeds, who earn a comfortable livelihood in fur-hunting and farming. It is 460 miles northwest of St. Paul, and 330 miles distant from Crow Wing. Notwithstanding the distance, there is considerable communication between the two places. West of Pembina, about thirty miles, is a settlement called *St. Joseph*, situated near a large mythological body of water called *Miniwakin*, or Devil's Lake.

"Now let me say something about this RED RIVER of the North, for it is beginning to be a great feature in this upper country. It runs north and empties into Lake Winnipeg, which connects with Hudson Bay by Nelson River. It is a muddy and sluggish stream, navigable to the mouth of the Sioux Wood River for vessels of three feet draught for four months in the year, so that the extent of its navigation within Minnesota alone (between Pembina and the mouth of Sioux Wood River) is 400 miles. Buffaloes still feed on its western banks. Its tributaries are numerous and copious, abounding with the choicest kind of game, and skirted with a various and beautiful foliage. It cannot be many years before this magnificent valley (together with the Saskatchewan) shall pour its products into our markets, and be the theatre of a busy and genial life.

"*Red River Settlement* is seventy miles north of Pembina, and lies on both sides of the river. Its population is estimated at 10,000 souls. It owes its origin and growth to the enterprise and success of the Hudson Bay Company. Many of the settlers came from Scotland, but the most were from Canada. They speak English and Canadian French. The English style of society is well kept up, whether we regard the Church with its bishop, the trader with his wine-cellar, the scholar with his library, the officer with his sinecure, or their paper currency. The great business of the settlement, of course, is the fur traffic.

"An immense amount of Buffalo skins is taken in summer and autumn, while in the winter smaller but more valuable furs are procured. The Indians also enlist in the hunts; and it is estimated that upward of $200,000 worth of furs are annually taken from our territory and sold to the Hudson Bay Company. It is high time indeed that a military post should be established somewhere on Red River by our government.

"The Hudson Bay Company is now a powerful monopoly. Not so magnificent and potent as the East India Company, it is still a powerful combination, showering opulence on its members, and reflecting a peculiar feature in the strength and grandeur of the British empire—a power which, to use the eloquent language of Daniel Webster, 'has dotted over the whole surface of the globe with her possessions and military posts, whose morning drum-beat following the sun, and keeping company with the hours, circles the earth daily with one continuous and unbroken strain of martial music.' The company is growing richer every year, and its jurisdiction and its lands will soon find an availability never dreamed of by its founders, unless, as may possibly happen, *popular sovereignty steps in to grasp the fruits* of its long apprenticeship."

The Charter of the Hudson Bay Company expired, by its own limitation, in 1860, and the question of annexing this vast domain to Canada, or forming a separate province, is now deeply agitating the British public, both in Canada and in the mother country.

TABLE OF DISTANCES,

From Fort William, SITUATED AT THE MOUTH OF THE KAMISTAQUOIAH RIVER, **to Fort Alexander,** AT THE HEAD OF LAKE WINNIPEG.

		Miles.
FORT WILLIAM..		0
Parapliue Portage..		25
(8 Portages)		
Dog Portage	51	76
(5 Portages)		
Savan or Swamp Portage*...	54	130
Thousand Islands Lake...	57	187
(2 Portages)		
Sturgeon Lake...	71	258
(4 Portages)		
Lac La Croix....	25	283
(5 Portages)		
Rainy Lake..	40	323
Rainy Lake River..	38	361
Lake of the Woods...	83	444
Rat Portage...	68	512
FORT ALEXANDER..	125	637

From Fort Alexander to For t Garry

OR RED RIVER SETTLEMENT, BY WATER.

		Miles.
To Pointe de Grand Marais..		24
" Red River Beacon...	25	49
" Lower Fort...	23	72
" FORT GARRY...	24	96

From FORT ALEXANDER to NORWAY HOUSE, passing through Lake Winnipeg, 300 miles.

From NORWAY HOUSE to YORK FACTORY, passing through Oxford Lake and Hayes River, 400 miles.

* Summit, elevated 840 feet above Lake Superior.

RAILROAD AND STEAMBOAT ROUTES

From Buffalo to Niagara Falls, Toronto, etc.

THE most usual mode of conveyance from Buffalo to the Falls of Niagara, and thence to Lake Ontario, or into Canada, is by the *Buffalo, Niagara Falls and Lewiston Railroad*, 28 miles in length. It runs through Tonawanda, 11 miles; Niagara Falls, 22 miles; Suspension Bridge, 24 miles, connecting with the Great Western Railway of Canada, and terminates at Lewiston, the head of navigation on Niagara River, 28 miles.

American and Canadian steamers of a large class leave Lewiston several times daily, for different ports on Lake Ontario and the St. Lawrence River.

There is also another very desirable mode of conveyance, by Steamboat, descending the Niagara River, from Buffalo to Chippewa, C. W., thence by the *Erie and Ontario Railroad*, 17 miles in length; passing in full view of the Falls, to the Clifton House, three miles below Chippewa; Suspension Bridge, five miles; Queenston, eleven miles, terminating at Niagara, C. W., thirty-five miles from Buffalo.

As the steamboat leaves Buffalo, on the latter route, a fine view may be obtained of Lake Erie and both shores of Niagara River. On the Canada side, the first objects of interest are the ruins of old FORT ERIE, captured by the Americans, July 3d, 1814. It is situated at the foot of the lake, opposite the site of a strong fortress which the United States government have recently erected for the protection of the river and the city of Buffalo.

WATERLOO, C. W., three miles below Buffalo and opposite Black Rock (now a part of Buffalo), with which it is connected by a steam-ferry, is handsomely situated on the west side of Niagara River, which is here about half a mile wide. The *Buffalo and Lake Huron Railroad* runs from Fort Erie, near Waterloo, to Paris, C. W., where it connects with the Great Western Railway of Canada. It is now completed to Goderich, C. W., lying on Lake Huron.

GRAND ISLAND, belonging to the United States, is passed on the right in descending the river. It is a large and valuable tract of good land, abounding with white oak of a superior quality.

NAVY ISLAND, belonging to the British, is next passed, lying within gun-shot of the mainland. This island obtained great notoriety in the fall and winter of 1837-'8, when it was occupied by the "Patriots," as they were styled, during the troubles in Canada. The Steamer *Caroline* was destroyed on the night of December 29th, 1837, while lying at Schlosser's Landing, on the American shore, having been engaged in transporting persons to and from the island, which was soon after evacuated.

Opposite Navy Island, on the Canada side, near Chippewa battle-ground, is the house in which Captain Usher resided, when murdered in 1838. It is supposed he fell by the hands of some of the deluded patriots, having been shot by a secret foe, while in his own house.

CHIPPEWA, 20 miles below Buffalo, and two miles above the Falls, is on the west side of Niagara River, at the mouth of a creek of the same name, which is naviga-ble to PORT ROBINSON, some eight or ten miles west; the latter place being on the line of the Welland Canal. The village of *Chippewa* contains a population of about 1,000 souls. Steamboats and lake craft of a large size are built at this place for the trade of Lake Erie and the Upper Lakes. It has obtained a place in history on account of the bloody battle which was fought near it in the war of 1812, between the United States and Great Britain. The battle was fought on the 5th of July, 1814, on the plains, a short distance south of the steamboat landing. The American forces were commanded by Major-General Jacob Brown, and the British, by Major-General Riall, who, af-ter an obstinate and sanguinary fight, was defeated, with considerable loss.

At Chippewa commences the railroad extending to Niagara, at the mouth of the river, a distance of 17 miles. Steamboats continue the line of travel from both ends of this road, thus furnishing an interesting and speedy conveyance between Lakes Erie and Ontario.

On ariving in the vicinity of the FALLS OF NIAGARA, the cars stop near the *Clifton House*, situated near the ferry leading to the American side. The site of this house was chosen as giving the best view of both the American and Canadian or Horse-Shoe Falls, which are seen from the piazzas and front windows. This is the most interest-ing approach to the Falls.

In addition to the Falls, there are other points of attraction on the Canada side of the river. The collection of curiosities at the Museum, and the Camera Obscura, which gives an exact and beautiful, though miniature image of the Falls, are well wor-thy of a visit. The *Burning Spring*, two miles above the Falls, is also much fre-quented; and the rides to the battle-grounds in this vicinity makes an exhila-rating and very pleasant excursion. For further description of Falls, see page 149.

DRUMMONDSVILLE, one mile west of the Falls, and situated on *Lundy's Lane*, is celebrated as the scene of another san-guinary engagement between the Ameri-can and British forces, July 25, 1814.

The following is a brief, though correct account of the engagement: "On the after-noon of the above day, while the Ameri-can army was on their march from *Fort George* toward *Fort Erie*, ascending the west bank of the river, their rear-guard, under the immediate command of Gen. Scott, was attacked by the advanced guard of the British army, under Gen. Riall, the British having been reinforced after their defeat at Chippewa, on the 5th of the same month. This brought on a general conflict of the most obstinate and deadly character. As soon as attacked, Gen. Scott advanced with his division, amounting to about 3,000 men, to the open ground facing the heights occupied by the main British army, where, were planted several heavy pieces of can-non. Between eight and nine o'clock in the evening, on the arrival of reinforcements to both armies, the battle became general and raged for several hours, with alternate success on both sides: each army evin-cing the most determined bravery and re-sistance. The command of the respective forces was now assumed by Major Gen. Brown and Lieut.-Gen. Drummond, each having under his command a well-disci-plined army. The brave (American) Col. Miller was ordered to advance and seize the artillery of the British, which he effected at the point of the bayonet in the most gallant manner. Gen. Riall, of the English army, was captured, and the pos-session of the battle-ground contested un-til near midnight, when 1,700 men being either killed or wounded, the conflicting armies, amounting altogether to about 6,000 strong, ceased the deadly conflict, and for a time the bloody field was left un-occupied, except by the dead and wounded.

When the British discovered that the Americans had encamped one or two miles

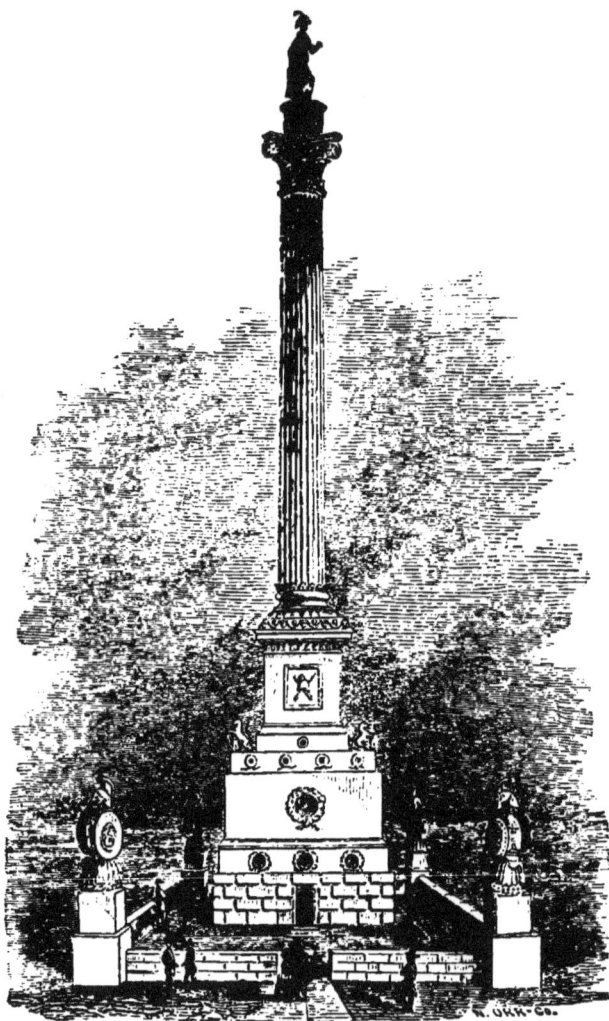

BROCK'S MONUMENT.—Queenston Heights.

distant, they returned and occupied their former position. Thus ended one of the most bloody conflicts that occurred during the last war; and while each party boasted a victory, altogether too dearly bought, neither was disposed to renew the conflict."

CLIFTON is a new and flourishing village, situated at the western termination of the Great Western Railway, where it connects with the *Suspension Bridge*. For description of route to Detroit, &c., see page 50.

QUEENSTON, situated seven miles below the Falls, and about the same distance above the entrance of Niagara River into Lake Ontario, lies directly opposite the village of Lewiston, with which it is connected by a Suspension Bridge 850 feet in length. It contains about 500 inhabitants, 60 dwelling-houses, one Episcopal, one Scotch Presbyterian, and one Baptist church, four taverns, four stores, and three warehouses. This place is also celebrated as being the scene of a deadly strife between the American and British forces, October 13, 1812. The American troops actually engaged in the fight were commanded by Gen. Solomon Van Rensselaer, and both the troops and their commander greatly distinguished themselves for their bravery, although ultimately overpowered by superior numbers. In attempting to regain their own side of the river many of the Americans perished; the whole loss in killed, wounded, and prisoners amounting to at least 1,000 men.

Major-General BROCK, the British commander, was killed in the middle of the fight, while leading on his men. A new monument stands on the heights, near where he fell, erected to his memory. The first monument was nearly destroyed by gunpowder, April 17, 1840; an infamous act, said to have been perpetrated by a person concerned in the insurrection of 1837-'38.

BROCK'S NEW MONUMENT was commenced in 1853, and finished in 1856;

being 185 feet high, ascended on the inside by a spiral staircase of 235 stone steps. The base is 40 feet square and 35 feet in height, surmounted by a tablet 35 feet high, with historical devices on the four sides. The main shaft, about 100 feet, is fluted and surmounted by a Corinthian capital, on which is placed a colossal figure of Major-General Brock, 18 feet in height. This beautiful structure cost £10,000 sterling, being entirely constructed of a cream-colored stone quarried in the vicinity. A massive stone wall, 80 feet square, adorned with military figures and trophies at the corners, 27 feet in height, surrounds the monument, leaving space for a grass-plot and walk on the inside of the enclosure.

The following is the inscription:

Upper Canada
Has dedicated this Monument
to the memory of the late
Major-General Sir ISAAC BROCK, K. B.
Provisional Lieut.-Governor and Commander
of the Forces in this Province,
Whose remains are deposited
in the vault beneath.
Opposing the invading enemy
He fell in action, near the Heights,
on the 13th October, 1812,
In the 43d year of his age,
Revered and lamented by the people
whom he governed, and deplored by
the Sovereign to whose service
His life had been devoted.

The last words of Major-General Brock, when he fell mortally wounded by a musket-shot through the left breast, were, "Never mind, my boys, the death of one man—I have not long to live." Thus departed one of the many noble spirits that were sacrificed on this frontier during the war of 1812.

The village of NIAGARA is advantageously situated on the Canada side, at the entrance of the river into Lake Ontario, directly opposite Fort Niagara, on the American side. It contains about 2,000 inhabitants, a court-house and jail; one Episcopal, one Presbyterian, one Methodist, and one Roman Catholic Church; 6 hotels and taverns; and 20 stores of different kinds; also, an extensive locomotive and car factory. This is the most noted place in Canada West for building steamboats and other craft navigating Lake Ontario. Here is a dockyard with a marine railway and foundry attached, capable of making machinery of the largest description, and giving employment to a great number of men. It is owned by the "Niagara Dock Company." Steamers leave daily for Toronto, etc.

Fort George, situated a short distance south or up-stream from the mouth of the river, is now in ruins. This was the scene of a severe contest in 1813, in which the Americans were victorious. A new fort has been erected on the point of land at the mouth of the river, directly opposite old *Fort Niagara* on the American side. The new fortification is called *Fort Massasauga.*

The whole frontier on the Canada side, from Fort George to Fort Erie, opposite Buffalo, was occupied by the American army in 1814, when occurred a succession of battles of the most determined and brilliant character.

NIAGARA RIVER,

ITS RAPIDS, FALLS, ISLANDS, AND ROMANTIC SCENERY.

" Majestic stream ! what river rivals thee,
Thou child of many lakes, and sire of one—
Lakes that claim kindred with the all-circling sea—
Large at thy birth as when thy race is run !
Against what great obstructions has thou won
Thine august way—the rock-formed mountain-plain
Has opened at thy bidding, and the steep
Bars not thy passage, for the ledge in vain
Stretches across the channel—thou dost leap
Sublimely down the height, and urge again
Thy rock-embattled course on to the distant main."

This most remarkable and romantic stream, the outlet of Lake Erie, through which flows all the accumulated waters of the Upper Lakes of North America, very appropriately forms the boundary between two great countries, the British province of Upper Canada on the one side, and the State of New York, the "Empire State" of the Union, on the opposite side. In its whole course, its peculiar character is quite in keeping with the stupendous Cataract from which its principal interest is derived.

The amount of water passing through this channel is immense; from a computation which has been made at the outlet of Lake Erie, the quantity thus discharged is about twenty millions of cubic feet, or upwards of 600,000 tons per minute, all of which great volume of water, 20 miles below, plunges over the Falls of Niagara.

The Niagara River commences at Bird Island, nearly opposite the mouth of Buffalo harbor, and passes by the site of old Fort Erie and Waterloo on the Canada side. At the later place a steam ferry-boat plies across the river to Black Rock, now forming a part of the city of Buffalo. It is here proposed to construct a railroad bridge across the stream, about 1,800 feet in width.

Squaw Island and Strawberry Island are both small islands lying on the American side of the stream, near the head of Grand Island. The river is here used in part for the Erie Canal, a pier extending from Squaw Island to Bird Island, forming a large basin called Black Rock Harbor.

Grand Island, attached to Erie Co.,

N. Y., is a large and important body of land, about ten miles long from north to south, and seven miles wide. This island is partly cleared and cultivated, while the larger portion is covered with a large growth of oaks and other forest trees.

The ship or steamboat channel runs along the bank of Grand Island to nearly opposite Chippewa, where the whole stream unites before plunging over the Falls of Niagara, being again separated at the head of Goat Island. From this point the awe-struck traveller can scan the quiet waters above, and the raging rapids below, preparing to plunge over the Cataract.

CAYUGA ISLAND and BUCKHORN ISLAND are small bodies of land belonging to the United States, situated immediately below Grand Island.

NAVY ISLAND, lying opposite the village of Chippewa, 18 miles below the head of the river, is a celebrated island belonging to the Canadians, having been taken possession of by the sympathizing patriots in 1837, when a partial rebellion occurred in Upper and Lower Canada.

TONAWANDA, 11 miles below Buffalo, is situated at the mouth of Tonawanda Creek, opposite Grand Island. The *Erie Canal* here enters the creek, which it follows for several miles on its course toward Lockport. A railroad also runs to Lockport, connecting with the *New York Central Railroad*, extending to Albany. A *ship canal* is proposed to be constructed from Tonawanda to some eligible point on Lake Ontario, thus forming a rival to the Welland Canal of Canada.

SCHLOSSER'S LANDING, two miles above Niagara Falls village, is a noted steamboat landing, opposite Chippewa, from whence the steamer *Caroline* was cut adrift by the British and destroyed, by being precipitated over the Falls during the Canadian rebellion, December 29th, 1837.

THE RAPIDS.—Below Navy Island, be-

tween Chippewa and Schlosser, the river is nearly three miles in width, but soon narrows to one mile, when the Rapids commence, and continue for about one mile before reaching the edge of the precipice at the Horse-Shoe Fall.

At the commencement of the Rapids, "the bed of the river declines, the channel contracts, numerous large rocks heave up the rolling surges, and dispute the passage of the now raging and foaming floods. The mighty torrent leaping down successive ledges, dashing over opposing elevations, hurled back by ridges, and repelled from shores and islands—plunging, boiling, roaring—seems a mad wilderness of waters striving against its better fate, and hurried on to destruction by its own blind and reckless impetuosity. Were there no cataract, these Rapids would yet make Niagara the wonder of the world."

IRIS, or GOAT ISLAND, commences near the head of the Rapids, and extends to the precipice, of which it forms a part, separating the American Fall from the Canadian or Horse-Shoe Fall. It is about half a mile in length, eighty rods wide, and contains over sixty acres of arable land, being for the most part covered with a heavy growth of forest trees of a variety of species, and native plants and flowers. A portion of the island, however, has been cleared off, and a garden enclosed, in which are some excellent fruit-trees, and a variety of native and foreign plants and flowers, and a fish-pond. The island is remarkably cool, shady, and pleasant, and is an object of unceasing admiration from year to year. Comfortable seats and arbors are placed at the most interesting points, where the visitor can sit at ease and enjoy the beautiful and sublime views presented to his sight—often entranced by a deafening roar of mighty waters in their descent, accompanied by changing rainbows of the most gorgeous description.

Niagara.

WRITTEN BY LYDIA H. SIGOURNEY.

Flow on forever, in thy glorious robe
Of terror and of beauty; God hath set
His rainbow on thy forehead, and the cloud
Mantles around thy feet, and He doth give
Thy voice of thunder power to speak of Him
Eternally; bidding the lip of man
Keep silence, and upon thy rocky altar
Pour incense of awe-struck praise.

GOAT ISLAND BRIDGE.—The Niagara Falls *Gazette* gives the following description of this new structure:

"This bridge across the east branch of the Niagara River is situated in the Rapids, about sixty rods above the Cataract, on the site of the old wooden bridge. It is 360 feet long, and consists of four arches of ninety feet span each, supported between the abutments of three piers. The piers above water are built of heavy cut stone, and are twenty-two feet long and six feet wide, tapering one foot in the height. The foundations are formed of foot-square oak timber, strongly framed and bolted together in cribs, filled with stone, and covered with timber at the surface of the water. These timber-foundations are protected against wear and injury from ice by heavy plates of iron, and being always covered with water, will be as durable as the stone.

"The superstructure is of iron, on the plan of Whipple's iron-arched bridge. The whole width is twenty-seven feet, affording a double carriage-way of sixteen and a half feet, and two foot-ways of five and a fourth feet each, with iron railings. The arches are of cast iron, and the chords, suspenders, and braces of wrought iron. All the materials used in the construction are of the best quality, and the size and strength of all the parts far beyond what are deemed necessary in bridges exposed to the severest tests.

"This substantial and beautiful structure, spanning a branch of this majestic river in the midst of the rapids, and overlooking the cataract, is worthy of the site it occupies, and affords another instance of the triumph of human ingenuity over the obstacles of nature.

"The islands connected by this bridge with the American shore are the property of Messrs. Porter, and constitute the most interesting features in the scenery surrounding the cataract. This bridge has been erected by them to facilitate communication with these interesting localities not otherwise accessible."

This is a toll-bridge, every foot passenger being charged 25 cents for the season, or single crossing.

There are upward of thirty islands and islets in the Niagara River or Strait, above the cataract. Most of those not described are small, and scarcely worthy of enumeration, although those immediately contiguous to Goat Island form beautiful objects in connection with the rushing and mighty waters by which they are surrounded. *Bath Island, Brig Island, Chapin's Island,* and *Bird Island*, all situated immediately above the American Fall, are reached by bridges.

When on Goat Island, turning to the right toward the Falls, the first object of interest is *Hogg's Back*, a point of land facing the American Fall,—Bridge to Addington Island immediately above the Cave of the Winds, 160 feet below. Sam. Patch's Point is next passed on the right, from which he took a fearful leap some years since. Biddle's Stairs descend to the water's edge below and the Cave of the Winds, which are annually visited by thousands of visitors. Terrapin Bridge and Terrapin Tower afford a grand view of the Canadian or Horse-Shoe Fall and Rapids above the Falls. Three Sister Islands are contiguous to Goat Island, on the American side. Passing around Goat Island toward the south, a grand view is afforded of the river and rapids above the Canadian and American Falls.

Niagara is a word of Indian origin—the orthography, accentuation, and meaning of which are variously given by different authors. It is highly probable that this diversity might be accounted for and explained by tracing the appellation through the dialects of the several tribes of aborigines who formerly inhabited the neighboring country. There is reason to believe, however, that the etymon belongs to the language of the Iroquois, and signifies the "*Thunder of Waters.*"

"When the traveller first arrives at the cataract he stands and gazes, and is lost in admiration. The mighty volume of water which forms the outlet of the great Lakes Superior, Michigan, Huron, and Erie, is here precipitated over a precipice 160 feet high, with a roar like that of thunder, which may be heard, in favorable circumstances, to the distance of fifteen miles, though, at times, the Falls may be nearly approached without perceiving much to indicate a tremendous cataract in the vicinity. In consequence of a bend in the river, the principal weight of water is thrown on the Canadian side, down

what is called the *Horse-Shoe Fall,* which name has become inappropriate, as the edges of the precipice have ceased to be a curve, and form a moderately acute angle. Near the middle of the fall, *Goat Island,* containing 75 acres, extends to the brow of the precipice, dividing the river into two parts; and a small projecting mass of rock at a little distance from it, toward the American shore, again divides the cataract on that side. Goat Island, at the lower end, presents a perpendicular mass of rocks, extending from the bottom to the top of the precipice. A bridge has been constructed from the American shore to Bath Island, and another connects the latter with Goat Island, and a tower is erected on the brow of the Horse-Shoe Fall, approached from Goat Island by a short bridge, on which the spectator seems to stand over the edge of the mighty cataract, and which affords a fine view of this part of it. The distance at the fall from the American shore to Goat Island is 65 rods; across the front of Goat Island is 78 rods; around the Horse-Shoe Fall, on the Canadian side, 144 rods; directly across the Horse-Shoe, 74 rods. The height of the fall near the American shore is 163 feet; near Goat Island, on the same side, 158 feet; near Goat Island, on the Canada side, 154 feet. Table Rock, a shelving projection on the Canadian side, at the edge of the precipice, is 150 feet high. This place is generally thought to present the finest view of the Falls; though, if the spectator will visit the tower on the opposite side on Goat Island, at sunrise, when the whole cavity is enlightened by the sun, and the gorgeous bow trembles in the rising spray, he cannot elsewhere, the world over, enjoy such an incomparable scene. A covered stairway on the American side descends from the top to the bottom of the precipice.

"It has been computed that 100 million tons of water are discharged over the precipice every hour. The Rapids commence about a mile above the Falls, and the water descends 57 feet before it arrives at the cataract. The view from the bridge to Goat Island, of the troubled water dashing tumultuously over the rocks of the American fall, is terrific. While curiosity constitutes an attribute of the human character, these falls will be frequented by admiring and delighted visitors as one of the grandest exhibitions in nature.

"This stupendous Cataract, situated in north latitude 43° 6', and west longitude 2° 6' from Washington, is 22 miles north from the efflux of the river at Lake Erie, and 14 miles south of its outlet into Lake Ontario. The whole length of the river is therefore 36 miles, its general course is a few points to the west of north. Though commonly called a river, this portion of the St. Lawrence is, more properly speaking, a *strait*, connecting, as above mentioned, the Lakes Erie and Ontario, and conducting the superfluous waters of the great seas and streams above, through a broad and divided, and afterward compressed, devious, and irregular channel to the latter lake, into which it empties—the point of union being about 40 miles from the western extremity of Lake Ontario.

"The climate of the Niagara is in the highest degree healthful and invigorating. The atmosphere, constantly acted upon by the rushing water, the noise, and the spray, is kept pure, refreshing, and salutary. There are no stagnant pools or marshes near to send abroad their fetid exhalations and noxious miasmas, poisoning the air and producing disease.

"Sweet-breathing herbs and beautiful wild flowers spring up spontaneously even on the sides, and in the crevices of the giant rocks; and luxuriant clusters of firs and other stately forest trees cover the islands, crown the cliffs, and overhang the banks of Niagara. Here are no mosquitoes to annoy, no reptiles to alarm, and no wild animals to intimidate, yet there is life and vivacity. The many-hued butterfly sips ambrosia from the fresh opened honey-cup; birds carol their lays of love among the spray-starred branches; and the lively squirrel skips chattering from tree to tree. Varieties of water-fowl, at certain seasons of the year, sport among the rapids, the sea-gull plays around the precipice, and the eagle—the banner bird of freedom—hovers above the cataract, plumes his gray pinions in its curling mists, and makes his home among the giant firs of its inaccessible islands.

"No place on the civilized earth offers such attractions and inducements to visitors as Niagara, and they can never be fully known except to those who see and study them, from the utter impossibility of describing such a scene as this wonderful cataract presents. When motion can be expressed by color, there will be some hope of imparting a faint idea of it; but until that can be done, Niagara must remain undescribed."

Cataract of Niagara.

"Shrine of Omnipotence! how vast, how grand,
How awful, yet how beautiful thou art!
Pillar'd around thy everlasting hills,
Robed in the drapery of descending floods,
Crowned by the rainbow, canopied by clouds
That roll in incense up from thy dread base,
Hid by their mantling o'er the vast abyss
Upon whose verge thou standest, whence ascends
The mighty anthem of thy Maker's praise,
Hymn'd in eternal *thunders!*"

Below the Falls, the first objects of interest are the Ferry Stairs and Point View on the American side; while on the opposite side is a ferry-house and landing, where carriages are usually to be found to convey passengers to the Clifton House, Table Rock, and other places of great interest.

About 30 rods below the Ferry Stairs is the spot where the hermit Abbot was drowned. Half a mile below the latter point is Catlin's Cave, formerly much frequented.

The SUSPENSION BRIDGE. the greatest artificial curiosity in America, is situated two miles and a half below the Falls, where has recently sprung into existence *Niagara City*, or better known as the *Suspension Bridge*, on the American side, and *Clifton* on the Canadian side of the river, here being about 800 feet in width, with perpendicular banks of 325 feet.

The *Whirlpool* and *Rapids*, one mile below the Bridge, are terrific sights of great interest, and well worthy a visit.

The *Devil's Hole*, one mile farther down, is also a point of great attraction, together with the *Bloody Run*, a small stream where a detachment of English soldiers were precipitated in their flight from an attack by Indians during the old French war in 1759. An amphitheatre of high ground spreads around and perfectly encloses the valley of the Devil's Hole, with the exception of a narrow ravine formed by Bloody Run—from which, against a large force, there is no escape, except over the precipice. The *Ice Cave* is another object of interest connected with the Devil's Hole.

The *Rapids* below the Whirlpool are the next object of attraction; then Queenston Heights and Brock's Monument on the Canadian side, and the *Suspension Bridge* at Lewiston; altogether forming objects of interest sufficient to fill a well-sized volume.

The Niagara River is navigable from Lewiston to its mouth at Fort Niagara, a farther distance of seven miles, or fourteen below the Falls of Niagara.

The village of NIAGARA FALLS, Niagara Co., N. Y., is situated on the east side of Niagara River, in the immediate vicinity of the grand Cataract, 22 miles from Buffalo and 303 miles from Albany by railroad route. No place in the Union exceeds this favored spot as a fashionable place of resort during the summer and fall months, when hundreds of visitors may be seen every day flocking to Goat Island, or points contiguous to the Rapids and Falls. The village contains several large hotels for the accommodation of visitors, the most noted of which are the Cataract House and the International Hotel; the Monteagle Hotel, situated two miles below the Falls, near the Suspension Bridge, and the Clifton House, on the Canada side, are all alike popular and well-kept hotels; there are five churches of different denominations; 15 stores, in many of which are kept for sale Indian curiosities and fancy work of different kinds. The water-power here afforded by the descending stream, east of Goat Island, is illimitable. A paper-mill, a flouring-mill, two saw-mills, a woollen factory, a furnace and machine shop, together with other manufacturing establishments, here use the water-power so bountifully supplied. The population is about 3,500.

The railroads centring at the Falls are the *Buffalo, Niagara Falls and Lewiston Railroad*, and the *New York Central Railroad*; the latter road connecting at Buffalo with the *New York and Erie Railroad*, and forming with other roads a direct route to Philadelphia, Baltimore, and Washington.

An *omnibus line* and hacks run from the village of Niagara Falls to Niagara City, or Suspension Bridge, during the summer months, and thence to the Clifton House and Table Rock on Canada side.

NIAGARA CITY, situated two miles below the Falls, at the *Suspension Bridge*, is a new and flourishing place containing about 1,500 inhabitants. Here is situated the *Monteagle Hotel*.

SUSPENSION BRIDGE

AND THE

Cataract and Rapids of Niagara.

To give you some idea of the grandeur of this triumph of engineering skill—THE SUSPENSION BRIDGE—we copy the following article from the Buffalo "Democracy" of June 21st, the character of which able and disinterested Journal requires no indorsement from us.

AN ENGINEER'S MONUMENT.

Spanning the chasm of the Niagara River, uniting the territories of two different Governments, and sustaining the uninterrupted railroad traffic of the Provinces of Canada with the United States, 250 feet above a flood of water which man has never been able to ferry, stands the monument of JOHN A. ROEBLING. The *Niagara Railway Suspension Bridge*, is the grandest and the most distinguishing achievement of Art in this world. It is the proudest, it is the most beautiful, and will prove to be the most enduring monument anywhere set up on this continent.

Regard this wonderful product of engineering skill. Its span is 822 feet. Yet an engine, tender and passenger car, loaded with men, and weighing altogether 47 tons, depress the long floor in the centre but 5½ inches. The Bridge, loaded with a loaded freight train, covering its whole length, and weighing 326 tons, is deflected in the middle only 10 inches. This extreme depression is perceptible only to practised eyes. The slighter changes of level require to be ascertained with instruments. Delicate as lace work, and seemingly light and airy, it hangs there high between heaven and the boiling flood below, more solid than the earthbeds of the adjacent railways. The concussions of fast moving trains are sensibly felt miles off through solid rocky soil. In cities locomotives shake entire blocks of stone dwellings. The waters of the Cayuga Lake tremble under the wheels of the express trains, a mile away from the bridge. But a freight train traversing JOHN A. ROEBLING'S Monument, at the speed of five miles an hour, communicates no jar to passengers walking upon the carriage way below. The land cables of the bridge do not tremble under it—the slight concussions of the superstructure do not go over the summits of the towers. This last fact in the stiffness of the great work is of much importance. It furnishes a guarantee of the durability of the masonry. Fast anchored with stone and grouted in solid rock cut down to the depth of twenty-five feet, the great cables are immovable by any mechanical force incidental to the use of the bridge, or the natural influences it will be subject to. The ultimate strength of these cables is 12,400 tons. The total weight of the material of the bridge, and of the traffic to which it will ordinarily be subjected is 2,262 tons, to sustain which the Engineer has provided in his beautiful and scientific structure, a strength of 12,400 tons. He demonstrates, too, that while the strength of the cables is nearly six times as great as their ordinary tension, THAT STRENGTH WILL NEVER BE IMPAIRED BY VIBRATION. This was the question raised by THE DEMOCRACY, a year ago, which excited such general, and in instances such angry discussion. ROEBLING treated our doubts with a cool reason and the stores of an extensive engineering experience, which gave us to believe that Art had at last attained to a method of suspending Iron Bridges for Railroad use, that should entirely obviate the objections to them felt by most of the Iron-Masters of the United States. He has since that demonstrated it in a most wonderful structure.

There are in the bridge 624 "suspenders," each capable of sustaining 30 tons—and all of sustaining 18,720 tons. The weight they have ordinarily to support is only 1,000 tons. But the Engineer has skilfully distributed the weight of the burdens, by the means of "girders" and "trusses." These spread the 34 tons heft of a locomotive and tender over a length of 200 feet. How ample is this provision made for defective iron or sudden strains!

The Anchor Chains are composed of 9 links, each 7 feet long, save -the last, which is 10 feet. The lowest link is made of 7 bars of iron, 7 inch by 1½. It is secured to a cast iron anchor plate 3½ inches thick, and 6 feet 6 inches square. The other links are equally strong. The iron used was all made from Pennsylvania charcoal, Ulster county, N. Y., and Salisbury Pig, and can be depended upon for a strength of 64,000 pounds to the square inch. The central portions of the anchor plates, through which the links pass is 12 inches thick. The excavations in the solid rock were not vertical. They inclined from the river. The rock upon which the work may rely on the New York side of the chasm is 100 feet long, 70 feet wide, and 20 feet deep. It weighs 160 pounds to the cubic foot, and presents a resistance of 14,000 tons, exclusive of the weight of the superincumbent masonry and embankment.

The Towers are each 15 feet square at the base, 60 feet high above the arch, and 8 feet square at the top. The limestone of which they are built will support a pressure of 500 tons on each square foot without crushing. While the greatest weight that can fall upon the tower will rarely exceed 600 tons, a pressure of 32,000 tons will be required to crush the top course. There are 4,000 tons' weight in each of the towers on the New York side.

The cables are 4 in number, 10 inches in diameter, and composed each of 3,640 small No. 9 wires. Sixty wires form one square inch of solid section, making the solid section of the entire cable 60.40 square inches, wrapping not included. These immense masses of wire are put together so that each individual wire performs its duty, and in a strain all work together. On this, Mr. ROEBLING, who is a moderate as well as a modest man, feels justified in speaking with the word PERFECT. Each of the large cables is composed of four smaller ones, called "strands." Each strand has 520 wires. One is placed in the centre. The rest are placed around that. These strands were manufactured nearly in the same position the cables now occupy. The preparatory labors, such as oiling, straightening, splicing, and reeling, were done in a long shed on the Canada side. Two strands were made at the same time, one for each of the two cables under process of construction. On the completion of one set, temporary wire bands were laid on, about nine inches apart, for the purpose of keeping the wires closely united, and securing their relative position. They were then lowered to occupy their permanent position in the cable, On completion of the seven pairs of strands, two platform carriages were mounted upon the cables, for laying on a continuous wrapping, by means of ROEBLING'S patent wrapping machines. During this process the whole mass of wire was again saturated with oil and paint, which, together with the wrapping, will protect them effectually against all oxidation. Five hundred tons of this wire is English. American manufacturers did not put in proposals. That used was remarkably uniform, and most carefully made.

The law deduced from large use of wire rope in Pennsylvania, is, that its durability depends upon its usage. It will last much longer under heavy strains moving slowly, than it will under light strains' moving rapidly. This law was borne constantly in mind by the Engineer of the Niagara Railway Bridge. The cables and suspenders are, so to speak, at rest. They are so well protected, too, from rust, that they may be regarded as eternally durable.

Among the interesting characteristics of this splendid architecture, is its elasticity. The depression under a load commences at the end, of course, and goes regularly across. After the passage of a train, the equilibrium is perfectly restored. The elasticity of the cables is fully equal to this task, and WILL NEVER BE LOST.

The equilibrium of the Bridge is less affected in cold weather than in warm. If a change of temperature of 100 degrees should take place, the difference in the level of the floor would be 2 feet 3 inches.

So solid is this Bridge in its weight, its stiffness, and its staying, that not the slightest motion is communicated to it by the severest gales of wind that blow up through the narrow gorge which it spans.

Next to violent winds, suspension bridge builders dread the trotting of cattle across their structures. Mr. ROEBLING says that a heavy train running 20 miles an hour across his Bridge, would do less injury to it than would 20 steers passing on a trot. It is the severest test, next to that of troops marching in time, to which bridges, iron or wooden, suspension or tubular, can be subjected. Strict regulations are enforced for the passage of hogs, horses, and oxen, in small bodies, and always on a walk.

This great work cost only $500,000. The same structure in England (if it could possibly have been built there) would have cost $4,000,000. It is unquestionably the most admirable work of art on this continent, and will make an imperishable monument to the memory of its Engineer, JOHN A. ROEBLING.

We append a Table of Quantities for the convenience of our readers, and the more easy comprehension of the character of the structure:

Length of span from centre to centre of
Towers........ .... 822 feet
Height of Tower above rock on American side............................. 88 feet
Height of Tower above rock, Canada side 78 feet
Height of Tower above floor of Railway 60 feet
Number of Wire Cables................ 4
Diameter of each Cable................ 10 inches
Number of No. 9 wires in each Cable.. 3,569
Ultimate aggregate strength of Cables, 12,400 tons
Weight of Superstructure............. 750 tons
Weight of Superstructure and maximum loads....................... 1,250 tons
Ultimate supporting strength..... ... 730 tons
Height of Track above water.......... 250 feet
Base of Towers.................. 16 feet square
Top of Towers... 8   "   "
Length of each Upper Cable........ 1,256¼ feet
"   "   Lower Cable........ 1,190 feet
Depth of Anchor Pits below surface of Rock............................ 30 feet
Number of Suspenders.............. 624
Ultimate strength of Suspenders...... 16,720 tons
Number of Overfloor Stays.......... 64
Aggregate strength of Stays........ 1,920 tons
Number of River Stays.............. 56
Aggregate strength of Stays......... 1,680 tons
Elevation of Railway Track above middle stage of River................... 245 feet
Total length of Wires.............. 4,000 miles

The weights of the materials in the bridge are as follows:

	LBS.
Timber...........................	919,180
Wrought Iron and Suspenders........	113,120
Castings.............................	44,332
Rails...............................	66,740
Cables (between towers)...............	535,400
Total........................	1,678,722

The GREAT WESTERN RAILWAY OF CANADA, which unites with the *New York Central Railroad*, terminating on the American side of the river, here commences and extends westward through Hamilton, London, and Chatham to Windsor, opposite Detroit, Mich., forming one of the great through lines of travel from Boston and New York to Detroit, Chicago, and the Far West. *See page 50.*

This road also furnishes a speedy route of travel to Toronto, Montreal, etc.

Rates of Charges at Niagara Falls.

The following are the rates of charges usually exacted from persons visiting Niagara Falls—but, unfortunately, impositions are often practised by unprincipled individuals at this, as well as other fashionable resorts:

AMERICAN SIDE.

Board, from one to two and a half dollars per day.

For services of guide, from one to three dollars.

For guide behind the Central Fall, and visiting the Cave of the Winds, one dollar.

For crossing bridge to Goat Island, 25 cents.

Fare to and from Suspension Bridge, 12½ cents.

Fare for crossing Suspension Bridge, 25 cents.

Fare to the Whirlpool, 50 cents.

For use of steps or cars on Inclined Plane, 5 cents.

Ferriage to Canada side, 20 cents.

Omnibus fare and steam ferriage to Canada side, 25 cents.

CANADA SIDE.

Board, from one to two and a half dollars per day.

Visiting Barnett's Museum, Camera Obscura, and Pleasure Grounds, 25 cents.

For guide and use of dress to pass behind the Fall at Table Rock, one dollar.

Carriage fare to Whirlpool, Lundy's Lane Battle Ground, Burning Spring, and back to Ferry, 50 to 75 cents.

Guide to Battle Ground and visiting Monument, 25 cents.

Carriage fare to Brock's Monument on Queenston Heights, one dollar.

Carriage fare per day, four dollars.

The drives in the vicinity of the Falls, on both sides of the river, are unrivalled, and no visitor should lose the opportunity to visit all the objects of attraction above and below the mighty Cataract.

It is necessary to make exact agreements with the hackmen and guides in order to avoid imposition; some on the Canada side refuse to take American bank-bills except at a great discount.

LEWISTON, Niagara Co., N. Y., is delightfully situated on the east bank of the Niagara River, seven miles below the Falls, and seven miles above the mouth of the river where it falls into Lake Ontario. It is an incorporated village, and contains about 1,000 inhabitants, four churches, an incorporated academy; a custom-house, it being the port of entry for the district of Niagara; three hotels, nine stores, and three storehouses. Here is a very convenient steamboat landing, from which steamers depart daily for Oswego, Ogdensburgh, etc., on the American side, and for Toronto, Kingston, etc., on the Canadian side. The Buffalo, Niagara Falls, and Lewiston Railroad terminates at this place, where is a magnificent Suspension Bridge thrown across the Niagara, connecting Lewiston with Queenston, Canada. The mountain ridge here rises about 300 feet above the river, forming many picturesque and romantic points of great interest. On the American side of the river stands the site of old Fort Gray, erected during the war of 1812, while on the Canadian side are situated Queenston Heights, surmounted by a beautiful monument erected to the memory of General Brock, of the British army, who was killed in a sanguinary conflict, October 13th, 1812. From this height a most extensive and grand view is obtained of Lake Ontario and the surrounding country.

YOUNGSTOWN, six miles below Lewiston, and one mile above old Fort Niagara at the mouth of the river, is a regular steamboat landing. The village contains about 800 inhabitants; three churches.

two public-houses, five stores, and two flouring mills, besides other manufacturing establishments. A railroad is nearly completed, extending from this place to Niagara Falls, being a continuation of the Canandaigua and Niagara Falls Railroad, now completed to the Suspension Bridge. A ferry plies from Youngstown to the village of Niagara on the Canada side of the river, here about half a mile in width. This is the first landing, on the American side of the river, after leaving the broad waters of Lake Ontario. *Fort Niagara* is situated at the mouth of the river.

Route around Lake Ontario.

	Miles.
Kingston. C. W., to Toronto, *via Grand Trunk Railway*	160
Toronto to Hamilton, C. W., *Toronto and Hamilton R. R.*	38
Hamilton to Suspension Bridge, *via Great Western R. R.*	43
Suspension Bridge to Rochester, N. Y., *via N. Y. Central Railway.*	76
Rochester to Oswego, N. Y., by *stage*	70
Oswego to Richland, N. Y., "	35
Richland to Cape Vincent, *via Watertown and Rome R. R.*	55
Cape Vincent to Kingston, C. W., *via Wolfe Island*	12

Total Miles	489

NOTE.—The extreme length of Lake Ontario is 190 miles, from Cape Vincent to Hamilton, C. W.; being about four times as long as its greatest width. The circuit of the water is estimated at 480 miles.— *See Lake Erie, page 14.*

LAKE ONTARIO.

This Lake, the most eastern of the great chain of Lakes of North America, receives the surplus waters of Niagara River; it is 180 miles in length, and 60 miles in extreme breadth; being about 480 miles in circumference. The boundary line between the British Possessions and the United States runs through the middle of the lake, and so continues down the St. Lawrence to the 45th degree of north latitude, where the river enters Canada.

The lake is navigable throughout its whole extent for vessels of the largest size; and it is said to be in some places upward of 600 feet in depth. Its surface is elevated 234 feet above the Atlantic, and lies 330 feet lower than Lake Erie, with which it is connected by the Niagara River and by the Welland Canal in Canada. It has also been proposed to construct a ship canal on the American side. The trade of Lake Ontario, from the great extent of inhabited country surrounding it, is very considerable, and is rapidly increasing. Many sail vessels and splendid steamers are employed in navigating its waters, which, owing to its great depth, never freeze, except at the sides, where the water is shallow; so that its navigation is not so effectually interrupted by ice as some of the other large lakes. The most important places on the Canadian or British side of Lake Ontario are Kingston, Coburg, Port Hope, Toronto, Hamilton, and Niagara; on the American shore, Cape Vincent, Sacket's Harbor, Oswego, Charlotte or Port Genesee, and Lewiston, on Niagara River. This Lake is connected with the navigable waters of the Hudson River by means of the Oswego and Erie canals. It receives numerous streams, both from the Canadian and the American sides, and abounds with a great variety of fish of an excellent flavor. The bass and salmon, in particular, have a high reputation, and are taken in large quantities. The principal Bays are Burlington, Irondequoit, Great and Little Sodus, Mexico, Black River, Chaumont, and the picturesque waters of the Bay of Quinte.

The passage across Lake Ontario in calm weather is most agreeable. At times both shores are hidden from view, when nothing can be seen from the deck of the vessel but an abyss of waters. The refractions which sometimes take place in summer, are exceedingly beautiful. Islands and trees appear turned upside down; and the white surf of the beach, translated aloft, seems like the smoke of artillery blazing away from a fort.*

* BEAUTIFUL MIRAGE.—That grand phenomenon occasionally witnessed on the Lake—mirage—was seen from the steamer Bay State, on a recent trip from Niagara to Genesee River (August, 1856), with more than ordinary splendor. The Lockport *Journal* says it occurred just as the sun was setting, at which time some twelve vessels were seen reflected on the horizon, in an inverted position, with a distinctness and vividness truly surprising. The atmosphere was overcast with a thick haze such as precedes a storm, and of a color favorable to represent upon the darkened background, vividly, the full outlines of the rigging, sails, etc., as perfect as if the ships themselves were actually transformed to the aerial canvas. The unusual phenomenon lasted until darkness put an end to the scene.

FORT NIAGARA—Mouth of Niagara River.

American Steamboat Route from Lewiston to Oswego, Kingston, and Ogdensburgh.

Ports, etc.	Miles.	Ports, etc.	Miles.
LEWISTON....................	0	OGDENSBURGH................	0
Youngstown..................	6	Morristown...................	11
Niagara, Can	1–7	Brockville, Can...............	1–12
Charlotte, or Port Genesee.......	80–87	Thousand Islands...........	
Pultneyville.................	20–107	Alexandria Bay..............	22–34
Sodus Point..................	10–117	Clayton, or French Creek.......	12–46
OSWEGO......................	30–147	Grand, or Wolfe Island........	
Stony Point and Island........	33–180	KINGSTON, Can...............	24–70
Sacket's Harbor..............	12–192	Sacket's Harbor..............	38–108
Grand, or Wolfe Island........	28–220	Stoney Point and Island.......	12–120
KINGSTON, Can...............	10–230	OSWEGO......................	33–153
Thousand Islands...........		Sodus Point.................	30–183
Clayton, or French Creek.......	24–254	Pultneyville................:	10–193
Alexandria Bay..............	12–266	Charlotte, or Port Genesee........	20–213
Brockville, Can..............	22–288	Niagara, Can.................	80–293
Morristown..................	1–289	Youngstown..................	1–294
OGDENSBURGH................	11–300	LEWISTON....................	6–300

USUAL TIME from Lewiston to Ogdensburgh, via Oswego and Kingston, 28 hours.
USUAL TIME, via Toronto and Cape Vincent, 22 hours.
Cabin Fare, $5.50 (including meals). Deck Fare, $2.50.

Steamboat Route from Lewiston to Toronto and Ogdensburgh, via Express Line.

Ports, etc.	Miles.	Ports, etc.	Miles.
LEWISTON	0	OGDENSBURGH...............	0
NIAGARA	7	Brockville, Can...............	11
TORONTO. Can...............	42–49	Clayton, or French Creek.......	34–45
Point Peter and Light.........128–177		CAPE VINCENT................	13–58
Duck Island..................	30–207	Tibbet's Point................	3–61
Tibbet's Point and Light........	19–226	Duck Island..................	19–80
CAPE VINCENT................	3–229	Point Peter and Light..........	30–110
Clayton, or French Creek.......	13–242	TORONTO 128–238	
Brockville, Can..............	34–276	NIAGARA	42–280
OGDENSBURGH................	11–287	LEWISTON...................	7–287

USUAL FARE from Ogdensburgh to Montreal, $3.50
Through Fare from Lewiston to Montreal,   9.00
"   "   from Buffalo to Montreal,   10.00

☞ For further information in regard to Lake Ontario and Route to Montreal, &c.,
see "PICTURESQUE TOURIST," published by J. DISTURNELL.

ALPHABETICAL LIST OF THE PRINCIPAL PORTS ON THE GREAT LAKES OF NORTH AMERICA, WITH THE SITUATION, TEMPERATURE, ETC.

PORTS, &c.	Latitude.	Longitude.	Altitude.	Mean Temp.
Agate Harbor, Mich,............	47°30′	88°10′	600 ft.	41° Fahr.
Algonac, "	42 36	82 30	570	46 00
Alpena, "	45 00	83 30	574	42 00
Amherstburg, Can...............	42 05	82 58	562	48 00
Ashland, Wis..................	46 33	91 00	600	41 00
Ashtabula, Ohio...............	41 52	80 47	560	47 00
Bay City, Mich.................			574	46 00
Bayfield, Wis..................	46 45	91 00	600	40 00
Beaver Bay, Min..............	47 12	91 18	600	40 00
Belleville, Can........			235	45 00
Brockville, "			230	44 00
Bruce Mines, Can..............	46 20	83 45	574	40 00
Buffalo,* N. Y.................	42 53	78 58	600	47 00
Cape Vincent, N. Y............	44 03	76 30	235	45 00
Charlotte, "	43 12	77 51	235	46 00
Chicago, Illinois...............	41 53	87 37	576	47 00
Clayton, N. Y..................	44 10	76 25	234	45 00
Cleveland,* Ohio...............	41 30	81 42	640	48 00
Cobourg, Can...................			235	45 00
Collingwood, Can..............	44 30	80 20	574	43 00
Conneaut, Ohio................			560	47 00
Copper Harbor,* Mich. (Ft. Wilkins)	47 30	88 00	620	41 00
Detroit,* "	42 20	83 00	600	47 25
Dunkirk, N. Y.................			569	47 25
Eagle Harbor, Mich............	47 28	88 18	600	41 00
Eagle River, "	47 25	88 30	600	41 00
East Saginaw, "			574	46 00
Erie,* Penn...................	42 08	80 05	560	47 00
Fairport, Ohio. .....·.........			560	47 00
Forrestville, Mich..............			574	45 00
Fort Gratiot,* "	42 55	82 23	598	46 30
Fort Niagara,* N. Y...........	43 18	79 08	250	47 90
Fort William, Can.............	48 23	89 22	600	36 00
Gena, Mich.,..................			576	43 00
Goderich, Can.................	43 44	81 43	574	45 00

PORTS, &c.	Latitude.	Longitude.	Altitude.	Mean Temp.
Grand Haven, Mich....;.........	43°05′	86°12′	576 ft.	46° Fahr.
Graud Portage, Min............	47 50	90 0C	600	38 00
Green Bay,* Wis., (Fort Howard)	44 30	88 05	620	44 50
Hamilton, Can.................			235	47 00
Hancock, Mich.................			600	41 00
Houghton, "	46 40	88 30	600	41 00
Huron Harbor, Ohio............			560	48 00
Kenosha, Wis.................	42 35	87 50	576	46 00
Kingston, Can.................	44 08	76 40	235	44 00
La Pointe, Wis................			600	40 00
Lexington, Mich...............			574	45 00
Lewiston, N. Y................			238	46 00
Manistee, Mich................			576	46 00
Manitouwoc, Wis..............	44 07	87 45	576	45 00
Mackinac,* Mich..............	45 51	84 33	728	40 65
Marquette, "	46 32	87 33	600	42 00
Michigan City, Ind............	41 50	87 06	576	49 00
Michipicoten, Can.............	47 56	85 06	600	38 00
Milwaukee, Wis...............	43 03	87 55	576	46 00
Monroe, Mich.................	41 53	83 19	560	48 00
Munising, "	46 20	87 00	600	41 00
Muskegon, Mich...............			576	46 00
Neepigon, Can.................	49 00	88 30	600	36 00
New Buffalo, Mich.............	41 45	86 46	576	47 00
Nenomonee City, Wis..........			576	43 00
Niagara, Can.................	43 18	79 08	235	47 00
Oconto, Wis..................			576	41 00
Oak Orchard, N. Y............			235	47 00
Ogdensburgh, N. Y............	44 42	75 35	230	44 00
Ontonagon, Mich..............	46 52	89 30	600	40 00
Oshawa, Can..................			235	44 00
Oswego.* N. Y., (Fort Ontario)...	43 20	76 40	250	46 44
Owen's Sound, Can............			574	43 00
Penetanquishene, Can..........	44 81	80 40	574	43 00
Picton, "			235	45 00
Port Burwell, "			560	46 00
Port Colburn, "			560	46 00
Port Dalhousio, "			235	47 00

PORTS, &c	Latitude.	Longitude.	Altitude.	Mean Temp.
Port Dover, Can................			560 ft.	46° Fahr.
Port Hope, "			235	45 00
Port Huron, Mich	42 58	82 25	572	46 00
Portland, Min..................	47 00	92 10	600	40 00
Port Stanley, Can..............			560	46 00
Prescott, "	44 42	75 36	230	44 00
Pultneyville, N. Y.............			235	46 00
Racine, Wis....................	42 45	87 48	576	47 00
Rock Harbor, Mich............	48 05	88 50	600	38 00
Sacket's Harbor,* (Madison Bar.)	43 55	76 00	265	45 00
Saginaw City, Mich............			574	46 00
Sandusky, Ohio................	41 27	82 45	560	48 00
Sarnia, Can...................	42 58	82 24	572	46 00
Saugeen, Can..................	44 04	81 43	574	44 00
Saut Ste. Marie,* (Fort Brady)....	46 30	84 43	600	40 37
Sheyboygan, Wis..............			576	45 00
St. Clair, Mich................			570	46 00
Superior, Wis..................	46 46	92 03	600	40 00
Sodus Bay, N. Y..............			265	46 40
Tawas, Mich..................			574	46 00
Toledo, Ohio...........	41 38	83 32	560	49 00
Toronto,† Can·................	43 40	79 20	265	44 40
Trenton, Mich................			566	47 00
Vermilion, Ohio...............			560	48 00
Waukegan, Ill.................	42 21	87 50	576	47 00
White River Harbor, Mich.......			576	46 30
Windsor, Can.................	42 21	83 00	570	47 00
Wyandotte, Mich..............			570	47 00

* United States Military Stations, giving the exact elevation of Forts, &c. The other Stations show the water level of the different Great Lakes and Rivers. :

† Canadian Observatory.

CEDAR RAPIDS.—St. Lawrence River.

1863.      1863.

Cleveland, Detroit, and Lake Superior.

The Splendid Steam Packets **METEOR** and **ILLINOIS** will leave Cleveland and Detroit for Lake Superior, as follows :

METEOR,	ILLINOIS,
R. S. RYDER, Master,	JOHN ROBERTSON, Master,
LEAVES CLEVELAND,	**LEAVES CLEVELAND,**
Wednesday, at 8 P. M.........July 8	Wednesday. at 8 P. M.........July 1
Tuesday, " .... " 21	Tuesday, " .... " 14
Monday, " ....Aug. 3	Monday, " .... " 27
Friday, " .... " 14	Friday, " ....Aug 7
Wednesday, " .... " 26	Wednesday, " .... " 19
Tuesday, " ....Sept. 8	Tuesday, " ....Sept. 1
Monday, " .... " 21	Monday, " .... " 14
Friday, " ....Oct. 2	Friday. " .... " 25
	Wednesday, " ....Oct. 7

Leaving Detroit on the day following those above named, at 10 A. M., calling at Port Huron and Sarnia the same evening.

During the months of July and August, the above Steamers will make

GRAND PLEASURE EXCURSIONS,

Leaving Cleveland on their regular days. On these trips they will carry good BRASS AND STRING BANDS, and every effort will be made to secure the comfort and convenience of passengers.   Each point of interest on the route will be visited, giving pleasure-seekers an opportunity to fully enjoy the finest, most healthy, and instructive trip on the Continent.

For further information, regarding Freight and Passage, address,

H. GARRETSON & CO., Agents,
No. 1 River Street, Cleveland, Ohio.

WILLIAMS & CO., Agents,
Foot of First Street, Detroit, Mich.

J. T. WHITING & CO.,
COMMISSION and INSURANCE AGENTS,
Also Agents for
LAKE SUPERIOR STEAMERS,
Foot of First Street, Detroit, Mich.

CLEVELAND, DETROIT, AND LAKE SUPERIOR LINE.

1863.

1863.

The First-Class Low Pressure Steamers **NORTHERN LIGHT** and **CITY OF CLEVELAND** will leave Cleveland for Lake Superior, regularly, on the days named below :

NORTHERN LIGHT,
JOHN SPALDING, Commander.

Monday,	at 8 P. M...........	July 6
Friday,	"	" 17
Wednesday,	"	" 29
Tuesday,	" ....Aug.	11
Monday,	"	" 24
Friday,	" ....Sept.	4
Wednesday,	"	" 16
Tuesday,	"	" 29
Monday,	"	Oct. 12

CITY OF CLEVELAND,
BENJAMIN WILKINS, Commander.

Friday,	at 8 P. M...........	July 10
Wednesday,	"	" 22
Tuesday,	" ....Aug.	4
Monday,	"	" 17
Friday,	"	" 28
Wednesday,	" ....Sept.	9
Tuesday,	"	" 22
Monday,	" ....Oct.	5

These Steamers will leave Detroit on the day following, at 10 A. M.

During the months of July and August, the above Steamers will make

GRAND PLEASURE EXCURSIONS,

Leaving Cleveland on their regular days. On these trips they will carry good BRASS AND STRING BANDS, and every effort will be made to secure the comfort and convenience of passengers. Each point of interest on the route will be visited, giving pleasure-seekers an opportunity to fully enjoy the finest, most healthy, and instructive trip on the Continent.

☞ Passengers will find their advantage in embarking for the trip at Cleveland, in having the first selection of rooms.

For further information, regarding Freight and Passage, address

WILLIAMS & CO., Agents Northern Light.

S. P. BRADY & CO., Agents City of Cleveland.

ROBERT HANNA & CO., Agents, Cleveland, Ohio.

1863.  1863.

FOR LAKE SUPERIOR.

The New and Splendid, Low Pressure, Side-wheel Passenger Steamboat

TRAVELLER,

F. S. MILLER, Commander,

Will leave Cleveland and Detroit for Ontonagon, touching at Sault Ste. Marie, Marquette, Portage Lake, Hancock, Houghton, Copper Harbor, Eagle Harbor, and Eagle River, on the days named below:

Leaves Cleveland at 8 P. M.	Leaves Detroit at 10 A. M.
Monday......................July 13	Wednesday ....July 1
Friday.......................July 24	Tuesday......................July 14
Wednesday....................Aug. 5	Saturday ....July 25
Tuesday ....Aug. 18	Thursday ....Aug. 6
Monday ....Aug. 31	Wednesday ....Aug. 19
Friday................✦......Sept. 11	Tuesday ....Sept. 1
Wednesday ....Sept. 23	Saturday ....Sept. 12
Tuesday ....Oct. 6	Thursday ....Sept. 24
	Wednesday ....Oct. 7

PLEASURE EXCURSIONS.

During the months of July and August, this boat will make **Four Grand Pleasure Excursions**, leaving Cleveland at 8 o'clock in the evenings of July 13th and 24th, and August 5th and 18th, and will leave Detroit on the mornings following her departure from Cleveland. To the tourist seeking health, pleasure, or valuable information, Lake Superior offers greater attractions than any other portion of the United States. The route embraces a thousand miles of diversified river and lake navigation, along the borders of which lies the most varied and grand scenery in the world, and no one can form any idea of the immense mineral resources of the country without a personal inspection of the vast iron and copper mines of this region. The distance up and back is about 2,000 miles, and occupies from nine to ten days. The boat stops long enough at each place to give passengers ample time to see all points of interest.

The price of Cabin Passage, including meals and berths, for the round trip to Ontonagon, returning on the same trip and boat, is, from Cleveland $35, and from Detroit $33. Servants, and children over three years old, half price. Children over twelve, full price.

For State Rooms, or further information, address

JOHN HUTCHINGS & CO.,
Foot of Griswold Street, Detroit.

1863.  1863.

LAKE SUPERIOR LINE.

The Splendid First-class Steamer **IRON CITY** will leave Cleveland and Detroit for Lake Superior, regularly, on days named below:

IRON CITY,

J. E. TURNER, Commander,

Leaves CLEVELAND, at 8 P. M.,

Friday ....July 3	Wednesday....................Sept. 2
Wednesday ....July 15	Tuesday ....Sept. 15
Tuesday......................July 28	Monday ....Sept. 28
Monday ....Aug. 10	Friday........................Oct. 9
Friday........................Aug. 21	

This Steamer will leave Detroit on the days following those named above, at 10 o'clock A. M.

During the summer months of July and August, the above Steamer will make

Five Grand Pleasure Excursions,

Visiting the different points of interest on Lake Superior, including the Pictured Rocks, and the various Copper and Iron Mines on its shores. For further information, and all particulars regarding Freight and Passage, address

S. P. BRADY & CO., Agents, Detroit, Mich.

HUSSEY & McBRIDE, Agents, Cleveland, Ohio.

Detroit and Cleveland

LINE OF  STEAMBOATS.

The Side-wheel Steamers,

MORNING STAR,	**MAY QUEEN,**
1,200 Tons,	700 Tons,
Capt. E. R. VIGER,	Capt. WM. M'KAY,

Leaving Cleveland and Detroit at Eight o'clock P. M.,

Form a DAILY EVENING LINE between Detroit and Cleveland, connecting with all early Morning Trains running East and West.

☞ Through Tickets for sale on board to all principal cities.

KEITH & CARTER, Agents, Detroit, Mich.

L. A. PIERCE, General Agent, Cleveland, Ohio.

Chicago Line.

FOR LAKE SUPERIOR,

The Splendid, First-Class Passenger Steamboat **PLANET**, Captain L. CHAMBERLIN, will run during the season of 1863, leaving Chicago at 7 o'clock in the Evening, for Ontonagon, Superior City, and all Intermediate Ports, on the following days :

Thursday, July 16, for Superior City.	Monday, September 28, for Ontonagon.
Monday, July 27, " "	Friday, October 9, for Superior City.
Tuesday, August 11, for Ontonagon.	Thursday, October 22, for Ontonagon.
Saturday, August 27, for Superior City.	Monday, November 2, for Superior City.
Friday, September 4, for Ontonagon.	Wednesday, Nov. 15, for Ontonagon.
Tuesday, Sept. 15, for Superior City.	

Her Dock is on River Street, first above Rush Street Bridge. For Freight **or** Passage, apply on board, or to

A. E. GOODRICH, 6 and 8 River Street.

Steamboats on Lake Michigan.

A First-Class Boat will leave Goodrich's Dock, first above Rush Street Bridge,

Every Morning (Sundays excepted),

At 9 o'clock, for

MILWAUKEE, KENOSHA, RACINE,

PORT WASHINGTON, SHEBOYGAN, MANITOWOC,

AND TWO RIVERS,

Extending their trips to Kewaunee and Wolf River every Friday. During the season of navigation, Passengers and Freight carried cheaper than by any other line.

Rates of Fare for Passengers.

	First Class.		Second Class.
Chicago to Kenosha	$1 00	....	$0 50
Chicago to Racine	1 25	....	75
Chicago to Milwaukee	1 50	....	1 00
Chicago to Port Washington	2 00	....	1 50
Chicago to Sheboygan	3 00	....	2 50
Chicago to Manitowoc and Two Rivers	3 50	....	3 00
Chicago to Grand Haven	3 00	....	2 50

☞ Passengers will please purchase their tickets on board the boats. First Class includes Meals and Berths. For Freight or Passage, apply on board, or to

A. E. GOODRICH, 6 and 8 River Street, CHICAGO

THE

NORTHERN TRANSPORTATION CO.

OF OHIO

Is prepared to Transport Property between

Boston, all Points in New England, New York, and the West,

With Promptness, Care, and Dispatch.

This well-known Line of Fifteen First-Class Screw Steamers

Connects at Ogdensburgh with the

Railroad for BOSTON and all Points in NEW ENGLAND;

At Cape Vincent with the

Railroads between Cape Vincent and New York;

And at Oswego with a

Line of Thirty First-Class Canal Boats between Oswego, Troy, Albany, and New York,

Form a Daily Line from

BOSTON, NEW YORK, OGDENSBURGH, CAPE VINCENT, AND OSWEGO TO CLEVELAND, TOLEDO, AND DETROIT,

And a Tri-weekly Line to

CHICAGO, MILWAUKEE, & INTERMEDIATE PORTS.

AGENTS.

J. Myers, 9 Astor House....New York.
Geo. A. Eddy..........Ogdensburgh.
John H. Crawford..........Oswego.
Walker & Hayes............Toledo.

John Hocking, 7 State Street..Boston
A. F. Smith............Cape Vincent.
Pelton & Breed..........Cleveland.
E. R. Mathews.............Detroit.

Grand Trunk Line of New Steamers.

B. F. WADE,	**MONTGOMERY,**
Capt. GOLDSMITH.	Capt. GILLIES.
ANTELOPE,	**WATER WITCH,**
Capt. BUTLIN.	Capt. RYDER.

The only reliable Line of Steamers from Chicago for Canada and the Eastern States having regular days and hours of sailing Tri-Weekly between

CHICAGO, MILWAUKEE, AND SARNIA.

One of the above Steamers will leave the Dock, foot of South La Salle Street,. Chicago, every

Tuesday, Thursday, and Saturday Evening,

At 7 o'clock ; and Milwaukee on

Wednesday, Friday, and Sunday Mornings,

At 7 o'clock, for SARNIA,

Landing at points on the West shore of Lake Michigan and Mackinac, connecting at Sarnia with the

GRAND TRUNK RAILWAY,

For Buffalo, Toronto, Oswego, Kingston, Prescott, Ottawa City, Montreal, Quebec, Portland,

And Eastern States. At Ogdensburgh with Northern New York and Vermont Central Railways, for St. Albans, Burlington, Montpelier, Concord, Lowell, Nashua, and all points in the New England States, forming a **Fast Freight Line** to all the above-named points.

ONLY ONE TRANSHIPMENT.

Rates of Insurance **Lower** than *via* any other route. Through Bills of Lading given to Liverpool, *via* Grand Trunk Railway and Montreal Ocean Steamships.

C. J. BRIDGES, Managing Director G. T. R., Montreal, C. E.
M. PENNINGTON, Freight Manager,   "   "   "
WILLIAM GRAHAM, Agent G. T. R., Portland.
GEORGE PHIPPEN, Agent G. T. R., No. 6 Devonshire Street, Boston.
S. T WEBSTER, Western Gen. Agent G. T. R., 56 Dearborn Street, Chicago, Ill.
A. T. SPENCER, Agent Grand Trunk Line Steamers, foot of South La Salle Street, Chicago, Ill.

H. COURTENAY, Agent,

Warehouse and Docks foot of Main Street, Milwaukee, Wis.

1863. 1863.

PLEASURE TRAVEL.

AMERICAN EXPRESS LINE.

Lake Ontario and River St. Lawrence.

Between Niagara Falls, Lewiston, Toronto, Ogdensburgh, Rouse's
Point, Montreal, Quebec, and River Saguenay.

**For Lake Champlain, Lake George, Saratoga Springs,
Troy, Albany, New York, White Mountains,
Portland, and Boston.**

The ONTARIO STEAMBOAT CO. will, during the season of Pleasure Travel,
commencing on the 22d of June, run their large and commodious Lake Steamers,

BAY STATE,	ONTARIO,	CATARACT,
Capt. MORLEY.	Capt. ESTES.	Capt. LEDYARD.

And the splendid River Steamers,

MONTREAL,	ALEXANDRA,
Capt. DEWITT.	Capt. J. N. BOCKUS.

Forming a Daily Line through Lake Ontario and River St. Lawrence.

LEAVE] DOWNWARD.	LEAVE] UPWARD.
TORONTO, daily (Sundays excepted) 6 30 A.M.	MONTREAL, daily, 7 00 A.M.
LEWISTON, daily, " " 10 30 "	OGDENSBURGH, daily (Sundays ex-
NIAGARA, daily, " " 10 50 "	cepted) 1 00 P.M.
CHARLOTTE, daily, " " 6 00 P.M.	PRESCOTT, daily (Sundays except'd) 1 10 "
OSWEGO, daily, " " 11 00 "	MORRISTOWN, daily, " " 2 00 "
For SACKET'S HARBOR, Thursdays and Sat-	BROCKVILLE, daily, " " 2 15 "
urdays, arriving next morning at 2 00 A M.	**Touching at Alexandria Bay & Clayton.**
KINGSTON, daily (Monday except'd)4 45 P.M.	KINGSTON, daily (Sunds. except'd) 10 00 P.M.
Touching at Clayton, Alexandria Bay,	SACKETS, daily, " " 1 30 A.M.
and Brockville.	OSWEGO, daily, " " 9 30 "
Arriving at OGDENSBURGH at 10 00 A.M.	CHARLOTTE, daily, " " 6 00 P.M.
And at MONTREAL same even'g, at 6 00 P.M.	Arriving at TORONTO at 5 00 A.M.

☞ This Line of Steamers is replete with all the comforts required by Travelers, and combines
the elegance of a First-class Hotel with the rapidity of Railroad conveyance. They are commanded and officered by men of experience, while the route offers to the Business man and Pleasure-
seeker attractions afforded by no other line, passing the far-famed THOUSAND ISLANDS BY
DAYLIGHT, at a time the most favorable for the grandeur of the Scenery and the comfort of the
Passengers. ☞ **Through Tickets** by this Line can be purchased at all points on the Line,
or on board the Steamers.

H. N. THROOP, General Manager, Oswego, N. Y.

SAMUEL FARWELL, President, Utica, N. Y.

Canadian Inland Steam Navigation
COMPANY.

Royal Mail **Through Line,**

For Darlington, Port Hope, Cobourg, Kingston, Brock-ville, Prescott, Ogdensburgh, and Montreal,

WITHOUT TRANSHIPMENT.

On and after MONDAY, the 4th of May,

One of the Steamers of the above Magnificent Line will leave the Custom House Wharf, foot of Yonge Street,

Daily (Sundays excepted), at 2 P. M.,

For the above PORTS. Also,

FOR HAMILTON every Morning, at 8 o'clock (Tuesdays excepted).

For Tickets and further information, apply at the Company's Offices, Front Street, adjoining the American Hotel, or the corner of York and Front Streets.

N. MILLOY, Agent.

TORONTO, May 14, 1863.

BAY SHORE ROUTE.

For Pensaukee, Oconto, Peshtigo, Marinette, and Sturgeon Bay.

The Steamer Queen City,

Captain J. A. MONROE,

Will run until further notice between Green Bay and the above-named Ports, leaving Green Bay on Monday, Wednesday, and Friday Mornings, at 7½ o'clock, and Marinette Tuesday, Thursday, and Saturday Mornings, at 6 o'clock. Will run into STURGEON BAY on her down trip every Friday.

JOHN B. JACOBS.

For Freight or Passage, apply on board, or to

STRONG & DAY, Agents, Green Bay.

GREEN BAY, *March* 31, 1863.

Milwaukee, Prairie du Chien & St. Paul
1863.   RAILWAY LINE.   1863.

**For Whitewater, Janesville, Monroe, Madison, Prairie du
Chien, McGregor, Winona, St. Paul, Jefferson, Fond
du Lac, Green Bay, Beloit, Freeport, Dunleith,**
And all Intermediate Points.

Trains leave Milwaukee immediately on arrival of Steamers of Detroit and Milwaukee R. R. Line, as follows :

10.20 A.M.—ST. PAUL'S EXPRESS, arriving at Prairie du Chien at 6.20 P.M., connecting with Steamers for St. Paul and Intermediate Points, arriving at St. Paul during the succeeding night, passengers remaining on board undisturbed until morning.

☞ This Train also makes direct connections for Beloit, Freeport, Rockford, etc.

10.20 P.M.—NIGHT EXPRESS, with Sleeping Car attached, arriving at Prairie du Chien at 9.40 A.M.

☞ Both the above Trains make direct connections at **Milton Junction** for Fort Atkinson, Jefferson, Fond du Lac, Oshkosh, Appleton, Green Bay, Berlin, etc., arriving at all these points at the same time as by competing routes.

Passengers for St. Paul and Intermediate Points, by taking this route, make the change from Cars to Steamers by daylight, obtaining Supper and a full night's rest on board, and arrive at St. Paul, etc., as soon as by any other route. These advantages can not be secured by any other route.

Through Tickets sold to all the above-named points, as low as by any other route.

FROM CHICAGO.
PASSENGERS FOR ST. PAUL

And Intermediate Points leave Chicago by Chicago and Northwestern Railway at 8.45 A.M.

Via Prairie du Chien,
Arrive at Prairie du Chien at 6.20 P.M.,

Making direct connection with Steamers, and getting Supper on board, and arrive at St Paul the succeeding evening.

☞ No change of Cars between Chicago and Prairie du Chien. No extra charge for Meals or State Rooms on Steamers.

The Splendid, First-Class Steamers,

Milwaukee, Key City, War Eagle, Itasca, and Northern Light,
Unequaled in elegance, speed, and comfort by any other Line,

LEAVE ST. PAUL DAILY, ABOUT 7 P. M.,

Arriving at Milwaukee at 3.50 P.M., and Chicago at 5.50 P.M., making direct connections at both points with Trains for the East.

WILLIAM JERVIS, Superintendent.

J C. SPENCER, General Manager.

Great Northwest Route

TO

ST. PAUL, MINN., AND LAKE SUPERIOR.

Direct Route to

Oshkosh, Fond du Lac, Berlin, Green Bay, on the North, and Prairie du Chien, La Crosse, St. Paul, etc., on the Northwest, via

Chicago and Northwestern Railway.

Cars run through to

JANESVILLE, WATERTOWN, FOND DU LAC, OSHKOSH, APPLETON, GREEN BAY, PRAIRIE DU CHIEN, LA CROSSE, Etc., without change.

Direct Route to

Rockford, Janesville, Madison, Freeport, Savanna, Galena, Dubuque, Beaver Dam, Portage, Kilbourn City, Berlin, Stevens' Point, Green Bay, Etc., Etc.,

And all points on the Mississippi River. ☞ Only Route without change of Cars.

THREE DAILY TRAINS leave Chicago—**8.45 A.M.** Day Express ; **5.00 P.M.** Janesville Accommodation ; **8.30 P.M.** Night Express, forming the direct and expeditious route to all points in the Northwest, connecting direct with **SPLENDID PACKETS** at Prairie du Chien and La Crosse, for all points on the Mississippi River. ☞ No charge on boats for Meals and State Rooms.

The Chicago and Northwestern Railway is now in splendid running order, and completely furnished with new and elegant

PATENT VENTILATED CARS,

Whereby the great annoyance of dust, so common on other roads, is avoided.

Superior arranged Sleeping Cars

Are run to Prairie du Chien, La Crosse, Fond du Lac, Oshkosh, and Green Bay.

☞ Passengers, to avail themselves of the many advantages of this Route over all others, should be particular and secure Tickets via "Chicago and Northwestern Railway."

☞ FARE ALWAYS AS LOW AS BY ANY OTHER ROUTE.

GEORGE L. DUNLAP, Superintendent.
E. DEWITT ROBINSON, General Ticket Agent.

Galena & Chicago Union
RAILROAD. .

The oldest and most reliable route to

THE NORTHWEST!

From CHICAGO to

Rockford, Warren, Galena, Freeport, Mineral Point, Dunleith, Dubuque,
Prairie du Chien, Lansing, Winona, Prescott, McGregor, La
Crosse, Reed's Landing, Hastings,

ST. PAUL!

**Beloit, Madison, Janesville, Pra. du Chien, Nottingham,
Waterloo, Independence, Cedar Falls.**

ILLINOIS, IOWA, AND NEBRASKA LINE

OF THE

GALENA AND CHICAGO UNION R.R.,

Consisting of Dixon and Fulton Air Line R.R., Chicago, Iowa, and Nebraska R.R.,
and Cedar Rapids and Missouri River R.R

From CHICAGO to

**Dixon, Fulton, De Witt, Toledo, Marshallt'n, Des Moines,
Council Bluffs, Sterling, Clinton, Cedar Rapids, In-
diantown, Boonsboro', Fort Dodge, Omaha City.**

Connecting with Stages for DENVER CITY, and all points in Western and Northern
Iowa and Nebraska.

No Change of Cars in Crossing the Mississippi River.

Connecting at **DUNLEITH** with Minnesota Packet Company's Daily Line of
Mail Steamers, during navigation, for **ST. PAUL.** And at Dubuque, with the
Dubuque and Pacific Railroad for Independence, Jesup, Cedar Falls, and all points
in Northern Iowa.

- **E. B. TALCOTT,** General Superintendent.
G. M. WHEELER, General Passenger Agent, CHICAGO.

Illinois Central Railroad.

THE GREAT THROUGH ROUTE

TO ST. LOUIS, PEORIA, SPRINGFIELD, DECATUR, CAIRO, MEMPHIS, via

ILLINOIS CENTRAL RAILROAD.

TWO EXPRESS TRAINS leave Chicago daily, on arrival of Trains from the East,

FOR

Jacksonville, Centralia, Decatur, Quincy, Peoria, Odin, Alton, St. Louis, Columbus, Leavenworth, St. Joseph, Springfield, Kankakee, Mattoon, Urbana, Tolono, Pana, Naples, Cairo, Memphis, Kansas City, Jefferson City,

And all parts of the South and Southwest.

BAGGAGE CHECKED THROUGH TO ALL IMPORTANT POINTS.

SLEEPING CARS WITH ALL NIGHT TRAINS.

PATENT DUSTERS ON DAY TRAINS.

Take Notice.—Memphis Passengers will find this the only direct route, and by purchasing tickets via I. C. R.R., will save distance, time, and money.

☞ Trains connect at Cairo daily with Steamers for Memphis.

Through Tickets for sale at the Office of the Company in the Great Central Dépôt, Chicago, also at all the principal Railroad Offices throughout the United States and Canada.

☞ Purchase Through Tickets via Illinois Central Railroad, and secure

Speed, Comfort, and Safety.

| W. P. JOHNSON, | W. R. ARTHUR, |
| General Passenger Agent, Chicago. | General Superintendent, Chicago. |

J. J. SPROULL,

General Agent, New York.

NEW YORK CENTRAL RAILROAD.

Connecting with Hudson River Railroad and Steamers.

For Buffalo, Niagara Falls, Detroit, Toledo,

CLEVELAND, CINCINNATI, CHICAGO,

Milwaukee, Madison, Rock Island, Iowa City, Dubuque, Burlington, Quincy, St. Paul, St. Louis, Cairo, &c.,

EITHER VIA

SUSPENSION BRIDGE, BUFFALO, OR NIAGARA FALLS,

Lake Shore Railroad, Buffalo and Lake Huron Railroad, or Great Western Railway (Canada).

THROUGH EXPRESS TRAINS

Leave Dépôt of **Hudson River Railroad,** Chambers and Warren Streets, New York, at 7 A.M., 10 A.M., and 5 P.M.

People's Line Steamers.

ISAAC NEWTON, HENDRICK HUDSON,

From foot of Courtlandt Street, every evening, at 6 P.M.

Passengers for Cleveland, Columbus, Cincinnati, Indianapolis, Terre Haute, Vincennes, Louisville, St. Louis, etc., can take **Lake Shore Railroad** from Buffalo or Niagara to Cleveland; thence by the Cleveland, Columbus, and Cincinnati Railroad, or the Cincinnati, Hamilton, and Dayton Railroad to above places, and all other points West and Southwest.

For **Through Tickets,** apply at the Office of the N. York Central Railroad,

239 Broadway, corner of Park Place, New York.

C. B. GREENOUGH, Passenger Agent.

JOHN H. MORE, Freight Agent.

SHERMAN HOUSE,

CHICAGO, ILLINOIS.

This HOTEL is centrally located on the corner of **Clark and Randolph S'reets,** pposite **Court House Square**; was built, in 1860, of Athens Marble, and has all the modern improvements, including a **Passenger Elevator** to convey the guests to and from the several stories of the house. In fact, it is in every particular, as

COMPLETE AND MAGNIFICENT AN ESTABLISHMENT

as there is in the United States.

DAVID A. GAGE,   }
                    } Proprietors.
CHARLES C. WAITE, }

TREMONT HOUSE,

CHICAGO, ILL.,

Situated on corner of Lake and Dearborn Streets.

Re-built, re-modeled, and re-furnished, in 1862, at a cost of $160,000   It contains all the modern improvements, and is one of the best-appointed Hotels in the country.   It has

NUMEROUS SUITES OF ROOMS,

with Baths, Water, etc., attached, for the accommodation of families.

It is easy of access to all the different Railroad Dépots, Places of Amusement, and Steamboat Landings.

GAGE & DRAKE, Proprietors

MISSION HOUSE,

MACKINAC, MICH.,

E. A. FRANKS, Proprietor.

This old and favorite Hotel is most delightfully situated on the romantic Island of Mackinac, within a short distance of the water's edge, and contiguous to the **Arched Rock, Sugar Loaf,** and other Natural Curiosities in which this famed Island abounds ; being alike celebrated for its pure air, romantic scenery, and fishing grounds.

Mackinac, *July*, 1862.

ISLAND HOUSE,

By Charles M. O'Malley,

MACKINAC, MICH.

The **ISLAND HOUSE** has been recently furnished throughout with New and Fashionable Furniture, and supplied with every facility to make it a First-class Hotel, and is

NOW OPEN FOR THE SEASON

for the entertainment of Travelers, Pleasure Parties, Invalids, and others, who desire a comfortable home while seeking pleasure or health in the pure atmosphere of Lake Superior, and the beautiful scenery of the surrounding country.

Mackinac, *July*, 1862.

McLEOD HOUSE,

MACKINAC, MICH.

This House is now open for the reception of guests. It has been repaired and enlarged, and furnished with entire new furniture. Its proximity to the Steamboat Landings, Places of Amusement, and business part of the town, makes it desirable for the Business Man and Pleasure-seeker, while the Invalid can rest under the **Extensive Piazzas,** and view the entire Town, Harbor, Fort, and Islands of the Straits, etc.

☞ An obliging Porter will be in attendance at the Boats to take charge of Baggage and conduct Passengers to the House.

R. McLEOD, Proprietor.

MACKINAC, *June* 18, 1862.

CHIPPEWA HOUSE,

SAUT STE MARIE,

MICHIGAN.

This favorite Hotel is pleasantly situated, near the Steamboat Landings, at the mouth of the **Ship Canal,** and in the immediate vicinity of Fort Brady.

No section of country exceeds the SAUT and its vicinity for

Fishing, Hunting, or Aquatic Sports.

The table of the Hotel is daily supplied with delightful White Fish and other varieties of the season, no pains being spared to make this house a comfortable home for the pleasure-traveler or man of business.

H. P. SMITH, Proprietor

TREMONT HOUSE,
MARQUETTE, MICHIGAN.

J. L. ARMSTRONG, Proprietor,

Begs leave to inform the Public that this Hotel is now open for the season. Having entirely REFURNISHED it throughout, and introduced all the modern improvements of a

FIRST-CLASS HOTEL,

He is now prepared to receive his guests in a manner unsurpassed by any other House on Lake Superior.

Having secured the services of **Mr. D. B. Hodges,** formerly of the Richmond House, Chicago, and the Massasoit House, Springfield, Mass., he hopes, with his assistance, to meet the approval of all who make the House a resort for

BUSINESS OR PLEASURE.

MARQUETTE, *June 1st,* 1862.

MARQUETTE HOUSE,
MARQUETTE, MICH.
L. D. JACKSON,
PROPRIETOR.

This favorite and well-kept House is

Delightfully Situated

near the Steamboat Landing, overlooking

LAKE SUPERIOR

and the adjacent country.

MASON HOUSE,
HANCOCK,
HOUGHTON COUNTY,
LAKE SUPERIOR,
S. C. SMITH,
PROPRIETOR.

This House is

PLEASANTLY SITUATED,

near the Steamboat Landing, overlooking

PORTAGE LAKE

and the adjoining country.

DOUGLASS HOUSE,

MR. BARSTOW,

PROPRIETOR,

HOUGHTON (Portage Lake),

MICHIGAN.

MICHIGAN EXCHANGE,

JEFFERSON AVENUE, DETROIT.
E. LYON, Proprietor.

INTERNATIONAL HOTEL,

IRA OSBORN, Superintend't.
NIAGARA FALLS, N. Y.

ST. LAWRENCE HALL,

GREAT ST. JAMES STREET,

MONTREAL.

This splendid HOTEL, which is situated in the most beautiful part of the City of Montreal, near the Banks and Post-Office, is furnished throughout in the best style of the New York and Boston Hotels, and comprises a

DINING SALOON AND CONCERT ROOM,

unequaled by any Hotel in Canada.

The TABLE will receive special attention, with the view of rendering it equal, if not superior, to any in America.

HOT AND COLD BATHS

can be had at all hours; and an **Omnibus** will always be in attendance on the arrival or departure of Railway Cars and Steamboats.

H. HOGAN & CO., Proprietors.

RUSSELL'S HOTEL,

PALACE STREET,

QUEBEC,

(UPPER TOWN.)

This well-managed and most comfortable HOTEL, kept by Messrs. RUSSELL, of Quebec, has recently been newly painted and re-furnished throughout. The Ball-room, used in summer, when the house is full of strangers, as a dining-room, has been entirely re-decorated in the handsomest style. The room will comfortably dine 250 persons at a time.—*Toronto Globe.*

COMMERCIAL
INSURANCE COMPANY,
MILWAUKEE,
WISCONSIN.
MARINE RISKS TAKEN AT CURRENT RATES.

Capital, $175,000, with a Surplus.

JOHN J. TALLMADGE, Pres.   G. D. NORRIS, V. Pres.   JAS. B. KELLOGG, Sec'y.

E. CRAMER, Treasurer.   L. H. LANE, Marine Inspector.

DIRECTORS.

F. Layton.	T. Littell.	D. Newhall.	J. Plankinton.	E. H Goodrich.
M B. Medbery.	Chas. F. Ilsley.	L H. Kellogg.	E. D. Chapin.	T. W. Goodrich.
M S. Scott.	G. D. Norris.	Robert Read.	L Sexton.	J. A. Dutcher.
D. Ferguson.	J. T. Bradford.	J. Bonnell.	G. Bremer.	E. Roddis.
O. E. Britt.	E. Cramer.	A L Hutchinson.	J. H. Inbusch.	S. T. Hooker.
Alex Mitchell.	William Young.	J. H. Cordes.	C. T. Bradley.	J. J. Tallmadge

---

CLEVELAND
IRON MINING COMPANY,
Miners and Dealers in
Lake Superior Iron Ore,
From their Iron Mountains, at Marquette, Lake Superior,
OFFICE AT CLEVELAND, OHIO.

DIRECTORS,

William J. Gordon. . . . . Cleveland.	George A. Tisdale. . . . . Cleveland.
John Outhwaite . . . . "	Samuel L. Mather . . . . "
George Worthington . . . . "	Isaac N. Judson . . . . New York.
S. D. McMillan . . . . "	M. L. Hewitt . . . . Marquette.

WILLIAM J. GORDON, President.

SAMUEL L. MATHER, Secretary and Treasurer

www.ingramcontent.com/pod-product-compliance
Lightning Source LLC
Chambersburg PA
CBHW021707210326
41599CB00013B/1558